RECYCLING AND REUSE OF MATERIALS AND THEIR PRODUCTS

Advances in Materials Science
Volume 3

RECYCLING AND REUSE OF MATERIALS AND THEIR PRODUCTS

Edited by

Yves Grohens, PhD, Kishor Kumar Sadasivuni, PhD
and Abderrahim Boudenne, PhD

Apple Academic Press
TORONTO NEW JERSEY

Apple Academic Press Inc.
3333 Mistwell Crescent
Oakville, ON L6L 0A2
Canada

Apple Academic Press Inc.
1613 Beaver Dam Road, Suite # 104
Point Pleasant, NJ 08742
USA

Exclusive worldwide distribution by CRC Press, a Taylor & Francis Group

International Standard Book Number: 978-1-926895-27-7 (Hardback)

Printed in the United States of America on acid-free paper

Library of Congress Control Number: 2012951942

Library and Archives Canada Cataloguing in Publication

Recycling and reuse of materials and their products/edited by Yves Grohens, Kishor Kumar Sadasivuni and Abderrahim Boudenne.

Includes bibliographical references and index.
ISBN 978-1-926895-27-7
1. Polymers--Recycling. I. Grohens, Yves II. Sadasivuni, Kishor Kumar, 1986- III. Boudenne, Abderrahim IV. Series: Advances in materials science (Apple Academic Press); v. 3

TP1175.R43R43 2013 668.9 C2012-906409-2

About the Editors

Yves Grohens, PhD

Prof. Yves Grohens (France) is the Director of the LIMATB (Material Engineering) Laboratory of Université de Bretagne Sud, France. His masters and PhD were from Besançon University, France. After finishing his studies he had been worked as assistant professor and later professor in various reputed Universities in France. He is an invited professor to many universities in different parts of the world as well. His areas of interest include physico chemical studies of polymers surfaces and interfaces, phase transitions in thin films, confinement, nano and bio composites design and characterisation and biodegradation of polymers and biomaterials. He has written several book chapters, monographs, scientific reviews and has 130 international publications. He is the chairman and member of advisory committees of many international conferences.

Director, LIMATB (Material Engineering) Laboratory, Centre de Recherche, rue St Maudé, Université de Bretagne Sud, Lorient 56100, France. Tel: & 00 33 2 97 87 45 08; Fax: 00 33 2 97 87 45 88. E-mail: yves.grohens@univ-ubs.fr Website: http://web.univ-ubs.fr/limatb

Kishor Kumar Sadasivuni, PhD

Dr. Kishor Kumar Sadasivuni has finished his Doctoral studies in Polymer nanocomposites. He has worked as a collaborative student between University of south Brittany (France) and Mahatma Gandhi University (India). He received his masters from the University of Andhra, India. He has about three years of experience in synthesis and characterization of nanoparticles and also in manufacturing elastomer nanocomposites. His areas of interest include different types of nanoparticles, their modifications, and their interactions with both modified and unmodified rubbers. He is also a visiting student of LECAP laboratory, Université de Rouen in France. He has presented papers at several international conferences worldwide.

Kishor Kumar S, Centre de Research, BP 92116, Université de Bretagne Sud, 56321 Lorient Cedex, France. Phone -0033297874581. Email: kishor-kumar.sadasivuni@univ-ubs.fr, chishorecumar@gmail.com

Abderrahim Boudenne, PhD

Dr. Abderrahim Boudenne (France) is a Professor of Material Science in Université Paris-Est Créteil, France. He has done his PhD in Energetics and Materials Science at Paris XII University in France. His post doctoral studies were entirely on Polymer Nanocomposites. His areas of expertise include: thermal, electrical, mechanical testing, numerical modeling of polymers and polymer composites materials. He is the chief editor of three books. He has several book chapters and many publications in international high impact factor journals in his credit. He presented papers in several international conferences. He organized many conferences and he is a reviewer for several international journals.

Dr. Abderrahim Boudenne, Professor of Polymer Science, Université Paris-Est Créteil, 61 Av. du Général de Gaulle 94010, Créteil France. Email: boudenne@u-pec.fr

About the Editors

[illegible]

Advances in Materials Science

Series Editors-in-Chief

Sabu Thomas, PhD

Dr. Sabu Thomas is the Director of the School of Chemical Sciences, Mahatma Gandhi University, Kottayam, India. He is also a full professor of polymer science and engineering and Director of the Centre for nanoscience and nanotechnology of the same university. He is a fellow of many professional bodies. Professor Thomas has authored or co-authored many papers in international peer-reviewed journals in the area of polymer processing. He has organized several international conferences and has more than 420 publications, 11 books, and two patents to his credit. He has been involved in a number of books both as author and editor. He is a reviewer to many international journals and has received many awards for his excellent work in polymer processing. His h Index is 42. Professor Thomas is listed as the 5th position in the list of Most Productive Researchers in India, in 2008.

Mathew Sebastian, MD

Dr. Mathew Sebastian has a degree in surgery (1976) with specialization in Ayurveda. He holds several diplomas in acupuncture, neural therapy (pain therapy), manual therapy, and vascular diseases. He was a missionary doctor in Mugana Hospital, Bukoba in Tansania, Africa (1976-1978) and underwent surgical training in different hospitals in Austria, Germany, and India for more than 10 years. Since 2000 he is the doctor in charge of the Ayurveda and Vein Clinic in Klagenfurt, Austria. At present he is a consultant surgeon at Privatclinic Maria Hilf, Klagenfurt. He is a member of the scientific advisory committee of the European Academy for Ayurveda, Birstein, Germany, and the TAM advisory committee (Traditional Asian Medicine, Sector Ayurveda) of the Austrian Ministry for Health, Vienna. He conducted an International Ayurveda Congress in Klagenfurt, Austria, in 2010. He has several publications to his name.

Anne George, MD

Anne George, MD, is the Director of the Institute for Holistic Medical Sciences, Kottayam, Kerala, India. She did her MBBS (Bachelor of Medicine, Bachelor of Surgery) at Trivandrum Medical College, University of Kerala, India. She acquired a DGO (Diploma in Obstetrics and Gynecology) from the University of Vienna, Austria; Diploma Acupuncture from the University of Vienna; and an MD from Kottayam Medical College, Mahatma Gandhi University, Kerala, India. She has organized several international conferences, is a fellow of the American Medical Society, and is a member of many international organizations. She has five publications to her name and has presented 25 papers.

Yang Weimin, PhD

Dr. Yang Weimin is the Taishan Scholar Professor of Quingdao University of Science and Technology in China. He is a full professor at the Beijing University of Chemical Technology and a fellow of many professional organizations. Professor Weimin has authored many papers in international peer-reviewed journals in the area of polymer processing. He has been contributed to a number of books as author and editor and acts as a reviewer to many international journals. In addition, he is a consultant to many polymer equipment manufacturers. He has also received numerous award for his work in polymer processing.

Advances in Materials Science

Series Editors-in-Chief

Contents

List of Contributors

Rameshwar Adhikari
Central Department of Chemistry, Tribhuvan University, Kirtipur, Kathmandu, Nepal.
Nepal Polymer Institute (NPI), P. O. Box–24411, Kathmandu, Nepal.

Srabanti Basu
Department of Biotechnology, Heritage Institute of Technology, Kolkata–700107, India.

Aparupa Bhattacharyya
Department of Chemical Engineering, National Institute of Technology, Durgapur, Durgapur–713209, India.

Edisley Martins Cabral
Universidade Federal do Amazonas; Address: Faculdade de Tecnologia-Av. General. Rodrigo O. J. Ramos, 3000, CEP 69077-000, Manaus-AM, Brazil.

Nilton Souza Campelo
Universidade Federal do Amazonas; Address: Faculdade de Tecnologia-Av. General. Rodrigo O. J. Ramos, 3000, CEP 69077-000, Manaus-AM, Brazil.

Soumasree Chatterjee
Research Fellow Department of Chemical Engineering National Institute of Technology, Durgapur Durgapur–713209, West Bengal, India.

P. Deepalekshmi
School of Chemical Science, Mahatma Gandhi University, Kottayam, Kerala–686560, India.

Susmita Dutta
Department of Chemical Engineering, National Institute of Technology, Durgapur, Durgapur–713209, India.

Y. Grohens
Limatb laboratory, Université de Bretagne Sud, Rue St Maudé–56100 Lorient, France.

G. C. Mohan Kumar
Department of Mechanical Engineering National Institute of Technology Karnataka, Surathkal, Mangalore–575025, India.

S. K. Kumar
LIMATB laboratory, Université de Bretagne Sud, Rue St Maudé, 56100 Lorient, France School of Chemical Sciences, Mahatma Gandhi University, Kottayam, Kerala, India.

Kamlesh Kumari
Department of Chemical Technology, Sant Longowal Institute of Engineering and Technology, Longowal Sangrur (Punjab) 148106, India.

Marko Likon
Insol Ltd., Cankarjeva 16a, 6230 Postojna, Slovenia.

Mirian Dayse Lima
Universidade Federal do Amazonas; Address: Faculdade de Tecnologia-Av. General. Rodrigo O. J. Ramos, 3000, CEP 69077-000, Manaus-AM, Brazil.

Joydeep Paul Majumder
Department of Civil Engineering Indian Institute of Technology, Kharagpur–721302, India.

R. B. Mane
Chemical Engineering and Process Development Division, National Chemical Laboratory, Pune 411008, India.

Anita Nair
H. O. D. Pulp and Paper Science Department and coordinator KUPG Centre, DES's Bangur nagar, Arts, Science and Commerce College Dandeli, Karnataka, India.

K. V. Pai
Professor of Industrial Chemistry, Kuvempu University, Jnana Sahydri, Shankaraghatta, Shimoga, Karnataka, India.

Dr. Anjali Pal
Department of Civil Engineering Indian Institute of Technology, Kharagpur–721302, India.

B. V. Reddy
Faculty of Engineering and applied Science University of Ontario Institute of Technology (UOIT) 2000 Simcoe Street North Oshawa, ON, Canada, L1H 7K4.

C. V. Rode
Chemical Engineering and Process Development Division, National Chemical Laboratory, Pune–411008, India.

Jouko Saarela
Syke Finnish Environment Institute, Mechelininkatu 34 a, Helsinki, Finland.

Bavan D. Saravana
Department of Mechanical Engineering National Institute of Technology Karnataka, Surathkal, Mangalore–575025, India.

Shobha Suryanarayan Sharma
Associate Professor of Chemistry, DES's Bangur nagar, Arts, Science and Commerce College Dandeli, Karnataka, India.

Claudia Cândido Silva
Universidade do Estado do Amazonas; Address: Escola Superior de Tecnologia-Av. Darcy Vargas 1200, CEP 69065-020, Manaus-AM, Brazil.

Virpal Singh
Department of Chemical Technology, Sant Longowal Institute of Engineering and Technology, Longowal Sangrur (Punjab) 148106, India.

T. Srinivas
School of Mechanical and Building Sciences Vellore Institute of Technology University (VIT U) Vellore, TN, India–632 014.

Sabu Thomas
School of Chemical Science, Mahatma Gandhi University, Kottayam, Kerala–686560, India.
Centre for Nanoscience and Nanotechnology, Mahatma Gandhi University, Kottayam, Kerala-686560, India.
UniversitiTeknologi MARA, Faculty of Applied Sciences, 40450 Shah AlamSelongor, Malaysia.
Center of Excellence for Polymer Materials and Technologies, Tehnoloski park 24, 1000 Ljubljana, Slovenia.

Raimundo Pereira de Vasconcelos
Universidade Federal do Amazonas; Address: Faculdade de Tecnologia-Av. General. Rodrigo O. J. Ramos, 3000, CEP 69077-000, Manaus-AM, Brazil.

Raimundo Kennedy Vieira
Universidade Federal do Amazonas; Address: Faculdade de Tecnologia-Av. General. Rodrigo O. J. Ramos, 3000, CEP 69077-000, Manaus-AM, Brazil.

Adalena Kennedy Vieira
Universidade Federal do Amazonas; Address: Faculdade de Tecnologia-Av. General. Rodrigo O. J. Ramos, 3000, CEP 69077-000, Manaus-AM, Brazil.

List of Abbreviations

ATSDR	Agency for Toxic Substances and Disease Registry
AvH	Alexander von humboldt
AIP	Alginate immobilized papain
AS	Anionic surfactants
ADM	Archer daniels midland
AAS	Atomic absorption spectrophotometer
AFM	Atomic force microscopy
AIBN	Azobisisobutyronitrile
BF	Bamboo fiber
BTNC	Bicarbonate treated tamarind nut carbon
BC	Bone charcoal
BR	Butadiene rubber
CEW	Carbonized material
CNSL	Cashew nut shell liquid
CNF	Cellulose nanofibers
CCD	Central composite design
CEPI	Central European Paranormal Investigations
CAUP	Chemically assisted ultrasound process
CHP	Combined heat and power
CMBR	Completely mixed batch reactor
CRD	Completely randomized design
COPE	Copolyesters
MDI	Diphenylmethanediisocyanate
DVB	Divinylbenzene
DHCP	Dow haltermann custom processing
ESCA	Electron spectroscopy for chemical analysis
EDS	Energy dispersive X-ray spectrometry
ENR	Epoxidized natural rubber
EG	Ethylene glycol
EPDM	Ethylene propylene diene monomer
EPM	Ethylene propylene monomer
EVA	Ethylene vinyl acetate
EIHA	European Hemp Association
EPP	Expanded polypropylene
EPR	Extended producer responsibility
FKM	Fluoroelastomer
FTIR	Fourier transforms infrared
FAPEAM	Fundação de Amparo a Pesquisa do Estado does Amazonas
GPC	Gel permeation chromatography
GHG	Greenhouse gas
GTR	Ground tire rubber

HDPE	High density polyethylene
HCW	Hot compressed water
IGCC	Integrated gasification combined cycle
IPA	Isopropanol
iPP	Isotactic polypropylene
LCA	Life cycle assesment
LLDPE	Linear low-density polyethylene
LR	Linear retraction
LL	Liquid limit
LDPE	Low-density polyethylene
LDI	Lysince-diiocyanate
MAPP	Maleic anhydride grafted polypropylene
MPP	Metal plated plastics
MFC	Microfibrillated cellulose
MSW	Municipal solid waste
NRP	Natural rubber powder
OPCC	Ordinary portland cement concrete
OSB	Oriented strand board
OEM	Original equipment manufacturer
1,2-PDO	1,2-Propanediol
PMS	Paper mill sludge
PI	Plasticity index
PVPi	Poly (vinyl pivalate)
PLA	Poly lactic acid
PBS	Polybutylene succinate
PAHs	Polycyclic aromatic hydrocarbon
PEBA	Polyether block amides
PE	Polyethylene
PET	Polyethylene terephthalate
PE/EVA	Polyethylene/ethylene-vinyl acetate copolymer
PHB	Polyhydroxybutyrate
PHBV	Polyhydroxybutyrate-co-polyhdroxyvalerate
PLA	Polylactic acid
PP	Polypropylene
PU	Polyurethane
PVA	Polyvinyl alcohol
PFA	Prevention of food adulteration act
PSOM	Pseudo second order model
RR	Reclaim rubber
NBRr	Recycled acrylonitrile-butadiene rubber
rHDPE	Recycled high density polyethylene
R-LDPE	Recycled low density polyethylene
RDF	Refuses derived fuel
RAFR	Relative air fuel ratio
RSM	Response surface methodology

RHA	Rice husk ash
SEM	Scanning electron microscopy
SEI	Secondary electron image
SCC	Self-compacting concrete
SCRC	Self-compacting rubberized concrete
SEA	Specific enzyme activity
SAIP	Spent alginate immobilized papain
SFR	Steam fuel ratio
SBR/NBRr	Styrene butadiene rubber/recycled acrylonitrile-butadiene rubber
SBR	Styrene butadiene rubber
SBCs	Styrenic block copolymers
SACC	Synthetic aggregate calcined clay
tBS	4-*tert*-Butylstyrene
TGA	Thermogravimetry anslysis
TPEs	Thermoplastic elastomers
TPNR	Thermoplastic natural rubber
TPO	Thermoplastic olefin
TPUs	Thermoplastic polyurethanes
TPVs	Thermoplastic vulcanizates
TOS	Time on stream
TDA	Tire derived aggregate
TDF	Tire Derived Fuel
TEM	Transmission electron microcopy
TOR	Trans-polyethylene rubber
WTD	Waste tyre dust
WTR	Waste tyre rubber
WA	Water absorption
WHO	World Health Organization

Preface

Polymers are becoming an inevitable part of day-to-day life. Their excellent properties make them applicable in all fields of subjects, especially in science and technology. Composites of polymers with micro- as well as nanoparticles made the scientific field available in one's hand. Various types of fillers such as nano inorganic compounds, carbon black, clay platelets and recently graphene, carbon nanotubes, etc., make polymers suitable in all areas, from automobile tires to space vehicles. But one of the biggest problem we are facing today is in connection with the disposal of wastes. In order to keep an eco-friendly atmosphere, all the polymer wastes should be carefully handled. The need of recycling and reuse of materials becomes significant in this regard. This book is an overall analysis of different methods and ways of recycling. It also deals with various types of polymers. A review chapter in this book is a study of all the work that had done in the past about this important aspect. All the chapters are highly informative and enlightening.

Chapter 1 deals with the effective conversion of biomass and municipal wastes to energy-generating systems as a measure to meet the growing demands of energy and also for reducing global warming. Co-generation systems are attracting much attention in this world of energy shortage. This technique as well as the co-firing technique used in biomass treatment now is discussed in detail here. The possibility of effectively utilizing the waste heat from industries and the role of exergy analysis in energy systems is nicely narrated.

Fossil fuels are being decreased in their amounts. So every where there is a need for biofuels. Production of biodiesel via transesterification of vegetable oils and animal fats has gained a tremendous thrust in recent years. This process releases much of the crude glycerol as a byproduct. Due to an increase in the production of crude glycerol and also due to the less refining capacity of it, the recycling of an excessive amount of crude glycerol for the production of commodity chemicals becomes an attractive option from both economical as well as ecological points of view. In Chapter 2, the authors tried to develop a catalyst for reuse of glycerol byproduct to a value -added commodity chemical, such as 1, 2-PDO. They found several catalysts such as barium, aluminium etc., in copper chromite. The metals act as promoters to the copper chromite catalyst. This chapter also explains the various methods of synthesis of chemicals from glycerol. The catalyst plays a major role in the reaction mechanisms of glycerol molecules.

Waste water contains much pollutant, and these causes serious environmental hazards. Waste water generated as a result of laundry and washing contains surfactants and detergents, which can adversely affect the soil and thus plant life. The surfactants can be removed from this water by adsorption since it has charge on it. This simple method of adsorption can be effectively utilized to produce the so-called treated waste water, which is also suitable for irrigation and so many other purposes where portable water is not required. Chapter 3 deals with these processes as well as the disposal of sludge in places where sand is used as both have same physical properties. This chapter gives a brief idea of a simple and comparatively cheap method of treating waste water to obtain eco-friendly products. The use of calcined clay to replace both fine and coarse aggregates as well as cement in concrete is discussed in the Chapter 4. Clay has been collected from different places, and it is calcined. The authors describe nicely the similarity in behavior of calcined clay and concrete. The method of production improved the workability without affecting the concrete specific gravity. This clay aggregates help to improve the compression strength to a very high value. Even though there is a technical advantage for clay calcined at high temperature, the economic advantage is less. The authors are also pointing out the need of further studies in this topic.

A considerable amount of research work has been done in the area of the recycling of rubbers as waste tire disposal is a serious problem in today's world. All aspects of waste rubber disposal issues and different methods of recycling are well described in Chapter 5. This chapter is a review and throws light on various research that has happened in the area so far. The authors not only explain the need and methods of recycling but also narrate the possible applications that recycled rubber components offer. The effective utilization of scrap and crumb rubbers in concrete and their blending with other polymers make them the most usable waste material. The chapter also analyses the environmental problems caused by both waste tires and its recycling process and points to a wide range of applications. Chapter 6 deals with the production of a sorbent material from paper mill sludge and replacing expanded polypropylene absorbent with this in oil spill sanitations. The chapter nicely narrates the process of production of absorbent from pulp and paper industry waste and how it is much more environmental friendly than synthetic ones. The one and only drawback in marketing the product is because of its slow degradation in water. The methods for eliminating this problem are also discussed in the chapter. The authors hope their product can be commercialized to a large extent in the near future.

Reuse of natural fibers for various industrial applications are discussed in Chapter 7. Natural fibers have the ability to replace synthetic ones. The authors critically explain that the manufacturers are not taking care in the recycling of wastes, but giving more emphasis to produce composites only. So the methods for recycling the composites are attracting more attention nowadays. The importance of cashew nut shell liquid as a source of surface active reagents is discussed in Chapter 8. The authors made an attempt to solve the problem of disposal of cashew nut shell liquid wastes and also an eco-friendly way to convert it into a useful product. As the authors indicated in the conclusion, this chapter would help in taking laboratory results to commercial utilization in the field of pulp and paper industry, as it includes the applications in the field of pulping, sizing, slime control, insecticide, and coating. Cadmium is a toxic waste disposed into the atmosphere as a byproduct of several industries. Removal of cadmium from various sources, its kinetics and equilibrium studies are becoming more important now. Chapter 9 demonstrates the absorption of cadmium from solutions by an enzyme. The chapter deals with the models for removal process and also gives an account of the cadmium recovery.

Chapter 10 deals with the integrated power and cooling systems based on biomass. the authors developed different options in integrated power systems and integrated combined power cooling systems based on biomass. Modeling and performance evaluation were done for various systems. They found that combined power and cooling integration results in higher fuel efficiency compared to the only power integrated systems. This chapter is very useful in today's world of growing energy demands. The disposal of nonbiodegradable heavy metals in industrial wastes has been a very significant problem in the technological point of view. Of the heavy metals, mercury plays an important role as a hazardous metal due to its wide distribution in nature and its toxicity to all forms of life. Chapter 11 presents several investigations carried out to prepare adsorbent from different types of waste materials to remove mercury from effluent as well as from drinking water. The chapter reviews the works carried out in the particular field of mercury removal, and it also deals with the challenges and future areas to be explored. The physiochemical properties of chitosan and starch particles in water are discussed in Chapter 12. The authors observed the viscosity variation even with very small amount of chitosan. For industrial applications such as fluid transport, this study is very useful.

Chapter 13 is a detailed study of the polymer composite with bamboo fibers. In this part the effect of surface treatments, modifications, and compatibilization effects of the fillers is described. One of the most useful fibers in nature is bamboo; its properties is given in the first part of the chapter followed by its composites with different polymers. The chapter critically reviews the mechanical property of the composites as well.

The last chapter is a general study on the eco-friendly green ways of recycling. Two important materials nanoparticles and polymers are considered in this section. Greener ways of synthesis and recycling help to avoid the toxic effects of conventionally used solvents and also avoid the emission of pollutants by consuming less energy. In the beginning of the chapters, the author addresses almost of the issues associated with the usual methods of material processing and later introduces the various green routes to solve the problems. As the final word an all recycling issues, this chapter matches both its role and purpose with the position of the book. This chapter really opens a greener door towards a safe future.

In total, this book aims to provide a basic understanding of the polymer recycling and reuse as well as the various methods for doing it. It also provides on a thorough knowledge of the work going on in different parts of the world about the recycling and also throws light on the areas to be explored in future.

— Yves Grohens, PhD, Kishor Kumar Sadasivuni, PhD
and Abderrahim Boudenne, PhD

1 Role of Waste Resources in Power Generation

B. V. Reddy and T. Srinivas

CONTENTS

1.1 INTRODUCTION

The world today is facing severe energy crisis due to industrialization and economic development on one hand and growing world population on the other. This growing energy demand is to be met in an environmental friendly manner with less pollution and reduced greenhouse gas emissions. Solar, wind, and biomass based technologies are being developed along with efficient and advanced energy technologies to effectively utilize the non renewable energy sources like coal, oil, and natural gas fuels. Biomass and municipal solid waste (MSW) based energy systems are receiving a great deal of attention to meet part of the growing energy demand because of their additional benefit in reducing global warming. Co-firing of biomass refuses derived fuel (RDF) in existing or in new coal fired power plants is an important option to reduce global warming. Cogeneration systems are notable due to their overall high energy utilization efficiency and reduced pollutants and greenhouse gas emissions. So based on the current global scenario, there is a need to focus on all possible energy sources to meet the energy requirement. This chapter presents details on biomass, MSW energy systems and cogeneration systems as well as its possible contribution to power generation. The biomass based energy systems and cogeneration systems will continue their importance in future as well since it offers high energy conversion efficiencies to meet the increasing global energy demand. The exergy analysis is receiving attention to analyze various energy systems to identify the sources of irreversibility. The exergy analysis and the benefits associated with exergy analysis are another topic of study of this chapter. The importance of biomass, MSW based co-firing power generation and cogeneration systems are also dealt with.

1.2 ADVANCED ENERGY SYSTEMS

1.2.1 Fluidized Bed Combustion Systems

The oil embargo in 1970s resulted in more research and development focus on solid fuels technologies. This resulted in the development of many advanced combustion technologies for industrial use and for power generation using low grade solid fuels. The fluidized bed combustion technologies and its next generation technologies such as circulating fluidized bed combustion, pressurized fluidized bed combustion are being employed for power generation. The fluidized bed combustion systems are flexible in burning coal, biomass, MSW, and other fuels. Coal, biomass, and MSW can be co-fired in fluidized bed combustion systems. There are many units in operation worldwide. Recently there is a great deal of interest on co-firing of coal and biomass in fluidized bed combustion systems to reduce the CO_2 emissions partly responsible for global warming. Coal, biomass gasification based combined cycle power generation systems are also developed to utilize coal with reduced greenhouse gas emissions.

1.2.2 Coal and Biomass Gasification Combined Cycle Power Generation

Coal, biomass, and other solid fuels can be gasified in a gasifier and the fuel gas is cleaned and is burnt in a gas turbine combustion chamber. The waste heat from the gas turbine exhaust is recovered in heat recovery steam generator to produce steam and to operate a steam cycle and this way the unit is operated in a combined cycle mode. There are power plants operating on gasification technology. A good amount of research is conducted on integrated gasification combined cycle (IGCC) systems to further improve their performance. There is also growing interest to develop IGCC systems for coal with carbon capture [1-3].

1.2.3 Co-firing of Coal, Biomass, and MSW in power generation systems

Co-firing of biomass, processed MSW with coal in steam power plants is receiving lot of attention in recent years. The coal and biomass can be fed into the boiler and biomass percentage can be adjusted to control heat input into the boiler. Already some power plants around the world employ this approach. The major benefit associated with co-firing of biomass, processed MSW in coal fired power plants is it reduces the carbon dioxide emissions from power plants. The other approach could be gasification of coal and biomass separately in gasification units and combustion of the combined syngas in a gas turbine combustion chamber to generate electric power using combined cycle technology. There is a good amount of research work in this direction and there is a growing interest in companies, research organizations to study the behavior of biomass, coal, MSW co-firing for different coals, biomass types and for different operating conditions. This is needed for proper operation of the units in co-firing mode.

1.2.4 Cogeneration Systems

Cogeneration is the simultaneous production of electric power and heat energy from the same fuel source. It also referred to as combined heat and power (CHP). The CHP systems can achieve very high overall energy conversions efficiencies. Cogeneration systems can result in a significant reduction in emissions including CO_2 [4]. Law and Reddy [5] investigated a natural gas fired combined cycle cogeneration system as shown in Figure 1. These units are becoming popular due to higher fuel utilization, higher conversion efficiencies, and reduced greenhouse gas emissions. The second law of thermodynamics and the exergy analysis will provide details on the role of combustor operating conditions, gas cycle and steam cycle operating conditions on the performance of the plant based on quality point of view. The exergy analysis as reported by Law and Reddy [5] for a natural gas fired combined cycle cogeneration system is presented in Figure 3. The exergy analysis presents the details on the performance of individual components from second law thermodynamics from quality point of view. The results demonstrate the role of exergy analysis in identifying the performance of components. The analysis also provides the methods for further improvement.

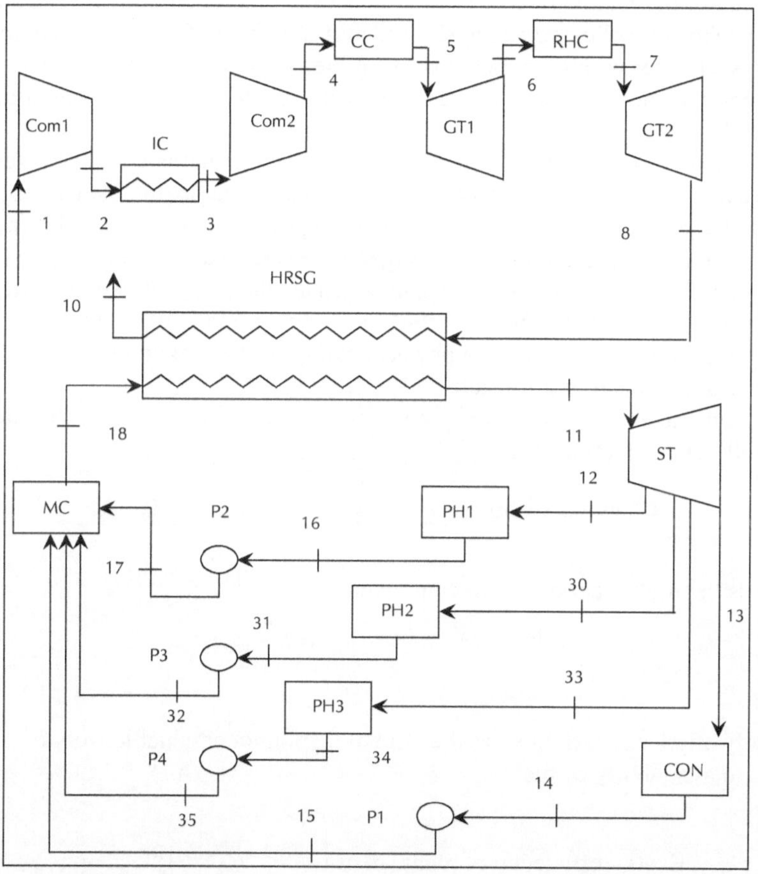

FIGURE 1 Schematic of combined cycle cogeneration unit with multiple process heaters [5].

1.3 ROLE OF EXERGY ANALYSIS ON ENERGY SYSTEMS

The exergy analysis is receiving a great attention from the last decade due to the ability to analyze a power generation system on a component basis and also as a whole system. The first law of thermodynamics presents energy balance for components or for the whole system. The second law provides insight into the performance of the energy system components and the whole energy system with quality point of view by analyzing the irreversibility in the components, losses in the components, and the performance of them with operating conditions. In the literature research investigation are conducted to improve the performance of the power generation system through exergy analysis leading to better design of system components and the whole power generation unit. Bejan [6] reported some details on exergy analysis. Rosen and Dincer [7] discussed the role of exergy analysis for thermal energy storage systems. Any improvement in the energy systems based on second law will result in greenhouse gas emissions leading to reduced global warming. The exergy analysis as reported by Law and Reddy [5] for a natural gas fired combined cycle cogeneration system is presented in Figure 3. The exergy analysis presents the details on the performance of individual components from second law thermodynamics from quality point of view. The results demonstrate the role of exergy analysis in identifying the performance of components. The analysis also provides the methods for further improvement. The advanced materials for combustion chambers and gas and turbine blades play a dominant role in the overall performance of a power generation system.

The total exergy of system is divided in to two parts that is chemical exergy and physical exergy. The chemical exergy of a fuel is maximum obtainable work by allowing the fuel to react with air from environment to produce environmental components of carbon dioxide, water vapor and nitrogen. In exergy balance of compressors and turbines, the chemical exergy cancels and so leaves the physical exergy which is the main contributor. However, chemical contribution plays main role in fuel gas combustor, reactor, gasifier, and so on. The physical exergy is the maximum obtainable work determined above the reference state.

The chemical and physical exergy components are determined at each state by using the following equations [8].

$$\text{Chemical exergy, } e_{ch} = \Sigma k\, n_k \varepsilon_k^0 + RT_0 \Sigma_k n_k \ln.[P.x_k] \tag{1}$$

where x_k is the mole fraction of *kth* component

$$\text{Physical exergy, } e_{ph} = h - \Sigma_k T_0 s_k \tag{2}$$

$$\text{Exergy, } e = e_{ch} + e_{ph} \tag{3}$$

The exergy efficiency is defined as the ratio of maximum obtainable work output from the plant to availability of fuel.

$$\text{Exergy efficiency of plant, } \eta_2 = \left(\frac{\varepsilon_{fuel}^0 - i_{total}}{\varepsilon_{fuel}^0} \right) \times 100 \tag{4}$$

Figure 4 presents the effects of typical biomass gasifier (as shown in Figure 2) characteristics on exergy efficiency of biomass gasifier for different biomass fuels. More details are presented in Srinivas et al. [9]. The exergy efficiency is high at lower relative air fuel ratio (RAFR) and steam fuel ratio (SFR) but it is low at low pressures. The exergy value of biomass fuel is different for different fuels. Therefore, there is a variation in exergy efficiency of gasifiers based on nature of fuel. The exergy efficiency of gasifier with rice husk is more than the other fuels. Manure exhibits a minimum exergy efficiency compared with the remaining fuels.

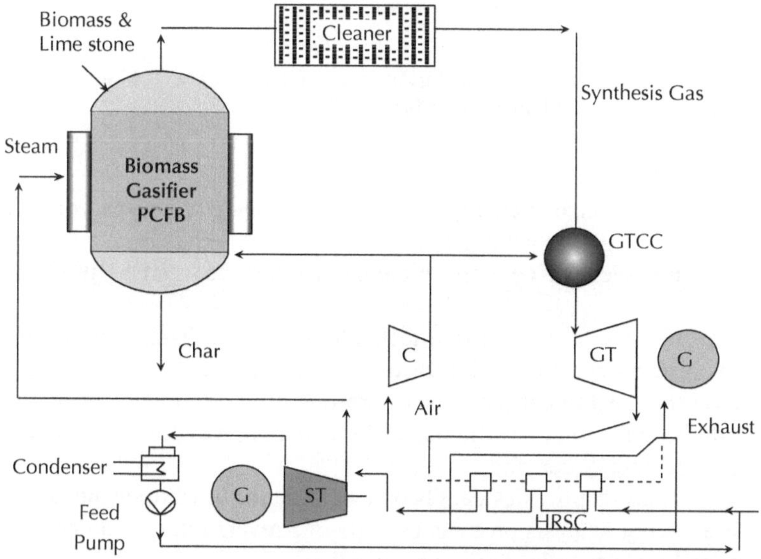

FIGURE 2 Schematic flow diagram of a biomass gasifier power generation system [9].

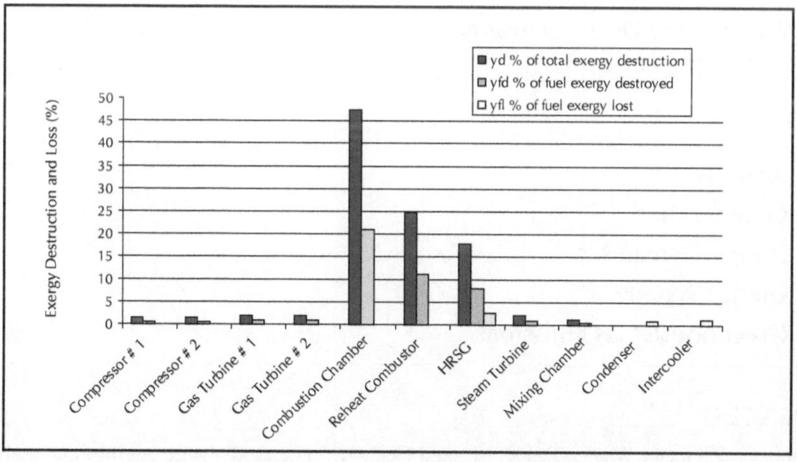

FIGURE 3 Exergy destruction and loss of all system components (TIT = 1600K, rp = 25) [5].

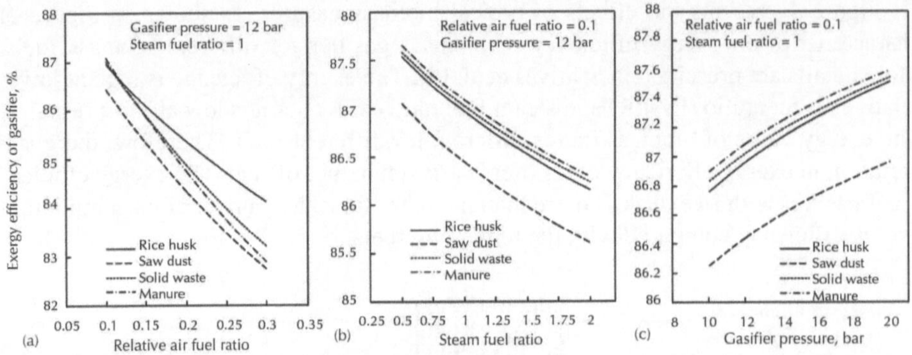

FIGURE 4 Exergy efficiency of biomass gasifier as a function of (a) RAFR, (b) SFR, and (c) gasifier pressure for different biomass fuels [9].

1.4 CONCLUSION

Biomass co-firing with coal in energy systems is one way to reduce part of greenhouse gases responsible for global warming. There is a need to develop advanced gasification based technologies for coal and biomass, so that they can be used for power generation.

With improvements in combustion technologies, the utilization efficiencies with cogeneration systems will further increase with time. There is a serious need to educate people on the need to conserve energy resources and to reduce energy wastage. Efforts should be made to utilize biomass, MSW in a big way for power generation to meet the growing energy demand.

The exergy analysis provides details on performance of components on a qualitative way. The exergy analysis presents the role of irreversibilities on performance of components and ways to reduce the irreversibility.

1.5 ACKNOWLEDGMENT

Dr. B. V. Reddy and Dr. T. Srinivas acknowledge the financial support from NSERC, Canada for the present work.

KEYWORDS

- **Biomass**
- **Cogeneration**
- **Energy Demand**
- **Energy Systems**
- **Greenhouse Gas Emissions**

REFERENCES

1. Wang, Z., Zhou, J., Wang, Q., Fan., J., and Cen., K. "Thermodynamic equilibrium analysis of hydrogen production by coal based on coal/CaO/H2O gasification system". *Int. J Hydrogen Energy*, 945-952 (2006).

2. Stiegel, G. J. and Ramezan, M. "Hydrogen from coal gasification: An economical pathway to sustainable energy future". *International Journal of Coal Geology*, **65**, 173-190 (2006).
3. Damen, K., Troost, M. V., Faaij, A., and Turkenburg, W. "A comparison of electricity and hydrogen production systems with CO_2 capture and storage. Part A: Review and selection of promising conversion and capture technologies". *Progress in Energy and Combustion Science*, **32**, 215-246 (2006).
4. Poullikkas, A. "An overview of current and future sustainable gas turbine technologies". *Renewable and Sustainable Energy Reviews*, **9**, 409-443 (2005).
5. Law, B. and Reddy, B. V. "*Energy and exergy analyses of a natural gas fired combined cycle cogeneration system*", Proceedings of ASME Energy Sustainability Conference, Paper No: ES2007:36257, Long Beach, CA, USA, (June 27–30, 2007).
6. Bejan, A. "Fundamentals of exergy analysis, entropy generation minimization, and the generation of flow architecture". *International Journal of Energy Research,* **26**, 545-565 (2002),
7. Rosen, M. A., Tang, R., and Dincer, I. "Effect of Stratification on Energy and Exergy Capacities in Thermal Storage Systems". *International Journal of Energy Research*, **8**, 177-193 (2004).
8. Kotas, T. J. "*The Exergy Method of Thermal Plant Analysis*". FL Krieger publishing company (1995).
9. Srinivas,T., Gupta, A. V. S. S. K. S., and Reddy, B. V. "Thermodynamic equilibrium model and exergy analysis of a biomass gasifier". *ASME Journal of Energy Resources Technology*, **131**(3), 1-7 (2009).

2 Catalytic Recycling of Glycerol Formed in Biodiesel Production

R. B. Mane and C. V. Rode

CONTENTS

2.1 INTRODUCTION

Increasing concern towards the sustainable society makes us imperative for recycle and reuse of materials which are either disposed of after their first use or which are co-generated in a chemical conversion along with the main product. The second category of materials/chemicals produced as by-products deserve special attention particularly, for developing new manufacturing processes due to growing concern for the atom efficiency and proper utilization of by-products. The commercial success sometimes even hinges for want of suitable recycle of by-products of the primary process. In the new era of developing non-conventional energy and chemical resources, biomass has proved to be a renewable, clean, carbon neural alternative for the conventional fossil based processes causing extensive global warming due to ever increasing demand for fuels and chemicals [1]. Biomass can be transformed in to fuel gas through partial combustion, biogas through fermentation, bioalcohol through biochemical processes, into a bio-oil or into a syngas and into most demandable biodiesel by transesterification of vegetable oils and animal tallow. The utilization of biomass helps to achieve two strategic goals of biorefinery that is (i) the displacement of imported petroleum in favor of renewable domestic raw materials (an energy goal) and (ii) the establishment of a robust biobased industry (an economic goal) and the closed loop as shown in Figure 1 [2].

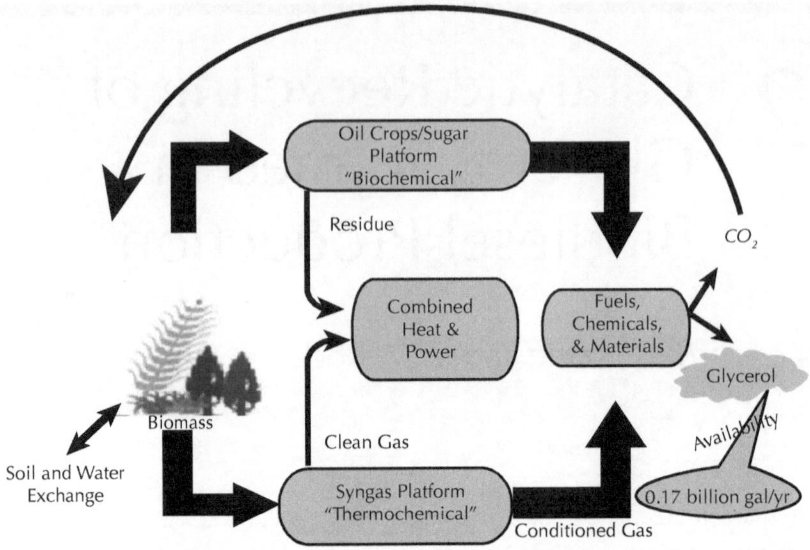

FIGURE 1 Biorefinery closed loop in the biomass and bioenergy chain.

In this regard, production of biodiesel *via* transesterification of vegetable oils and animal fats has gained a tremendous thrust in recent years. Transesterification process yields crude glycerol as a by-product, at the rate of 10% of per unit of biodiesel having 85% glycerol (range: 40–90%), 10% water (range 8–50%), 4% salt (range 0–10%), less than 0.5% methanol, and around 0.5% free fatty acids and it can be separated from the crude mixture by gravity, neutralization with caustic soda and distillation for removal of methanol and water [3]. Nowadays, worldwide biodiesel production has tremendously increased, for example in 2004 there were 22 plants having biodiesel production capacity of 157 million gal/year while today 148 plants are producing biodiesel at the rate of 1.4 billion gal/year and it will reach to 8.2 billion gallons by 2020. Consequently this will result in 820 million gallons of glycerol entering in the market from biodiesel only and leads to decrease in glycerol market price as shown in Figure 2 [4]. The average price of crude glycerol is stable around 140–$180/ton in 2010 [2] and for crude glycerol it is about $0.11 per kg. Glycerol is also obtained in large quantities from soap splitting industries mainly engaged in the production of health care products [5].

As per the available information, in United States, very few companies refine glycerol; largest producers include Proctor & Gamble (150 million pound), Dial, and Cognis, which are all major players in the global chemical market. The total refining capacity of crude glycerol is barely >500 million ponds/year which is not enough to refine the glut of crude glycerol, now becoming available through various routes. Although, refined glycerol has about 1,500 uses out of which 64% is used in food products, personal care products, and oral care, these are either diminutive volume applications or having a small market volume due to its higher production cost compared to crude glycerol. Hence, the recycle of excessive amount of crude glycerol for the production of commodity chemicals becomes an attractive option from both economical as well as ecological points of view. As the glycerol is highly functionalized molecule, it undergoes various transformations such as dehydration, hydrogenolysis, oxidation, etherification, and esterification (Scheme 1). Among various options, selective catalytic hydrogenolysis of glycerol to 1,2-propanediol (1,2-PDO) is an imperative process as 1,2-PDO is being produced over 0.5 MMTPA with overall growth rate of 4% [6, 7].

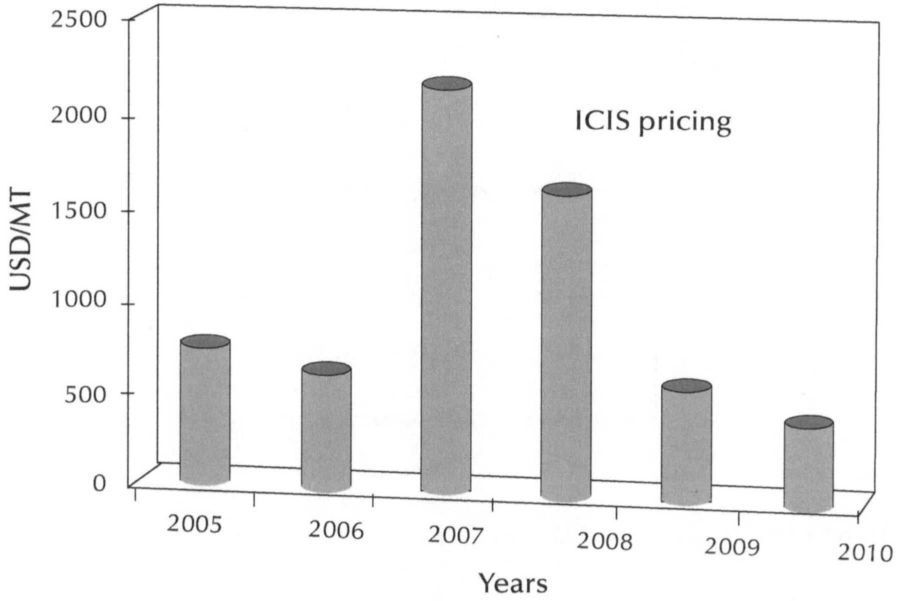

FIGURE 2 Price history of refined glycerol.

A global demand of 1,2-PDO is estimated to rise to ~1–1.5 MMTPA because of its wide spread applications in unsaturated polyesters required in fiber glass reinforced structures, surface coatings, and paints (Figure 3) and also because of its lower toxicity preferred over ethylene glycol [8]. Commercially 1,2-PDO was produced by the hydration of propylene oxide derived from propylene either by chlorohydrin process or hydroperoxide process (Scheme 2).

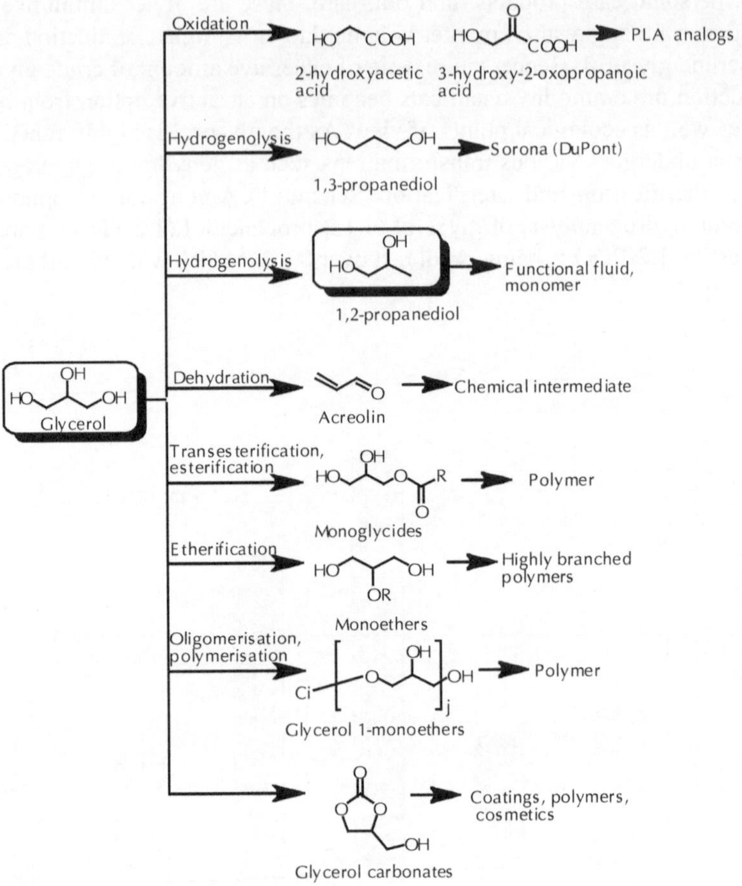

SCHEME 1 Downstream applications of glycerol

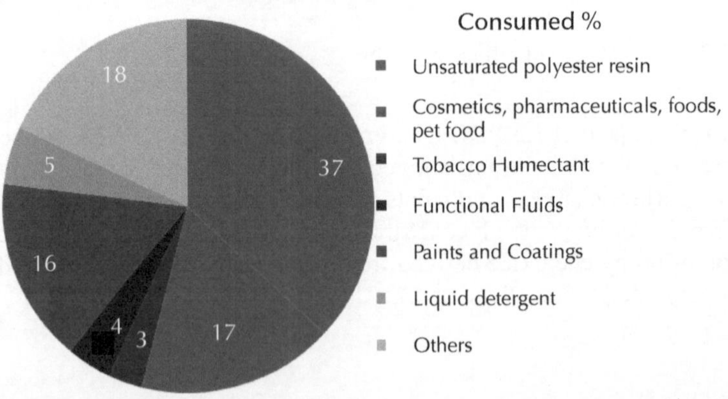

FIGURE 3 Uses of 1,2-propanediol.

This process has drawbacks of (i) lowering petroleum sources (ii) sensitive to crude prices (iii) liberation of hydrochloric acid.

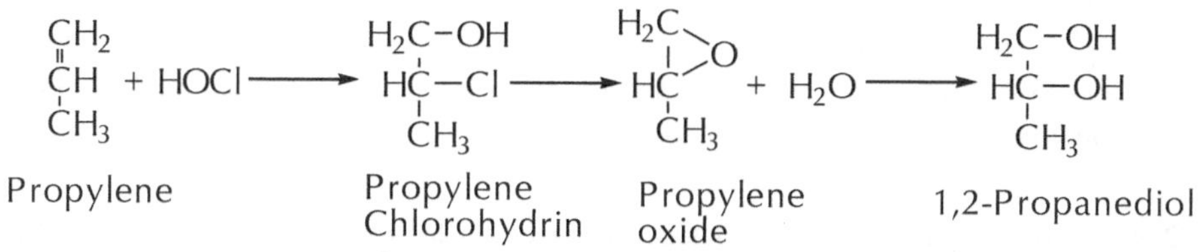

SCHEME 2 Commercial chlorohydrins process for production of 1,2-PDO.

Therefore, use of renewable glycerol for the production of 1,2-PDO provides sustainable process by reducing the carbon foot prints and offer economical benefits making biodiesel production a cost effective process.

2.2 CURRENT SCENARIO

Considering the surplus availability of glycerol and increasing demand of 1,2-PDO due to its wide spread applications in different sectors, several researchers, and industries have focused their research interests on catalytic conversion of glycerol to 1,2-PDO. There are several reports and/or papers appearing on catalytic hydrogenolysis of glycerol and some important findings are summarized here. Catalysts reported for glycerol hydrogenolysis mainly include mono, bi metallic noble metals (Rh, Ru, Pt, Pd) and/or transition metals, mainly Cu, Ni, Co, Cr, Mg, and Zn, supported on porous materials (usually activated carbon, SiO_2 or Al_2O_3). Hydrogenolysis was performed either with pure glycerol solution, or usually in water and also in other solvents such as alcohols and sulfolanes, and so on. Among the noble metal catalysts Rh/SiO_2, and TiO_2 supported Ru catalysts showed higher glycerol conversion in the range of 20–90% under mild reaction conditions but showed lesser 1,2-PDO selectivity (20–40%) [9, 10]. Supported noble metal catalysts in combination with acid materials such as amberlyst, niobia, 12-tungstophosphoric acid supported on zirconia, and heteropoly acid salt Cs 2.5H0.5[$PW_{12}O_{40}$] were also studied [11-14]. They found linear correlation between conversion and acidity of the catalysts. The promotion effect of Re on catalysts performance, in terms of both glycerol conversion and 1,2-PDO selectivity was also investigated [15].

Transition metal catalysts mainly include supported as well as mono, bi metallic Cu, Mg, Co, Ni, Cr, and Ag which were reported to exhibit good selectivity towards 1,2-PDO [16-22]. Among these catalysts systems, only copper chromite has been considered for commercialization due to its better performance under mild reaction conditions. In spite of extensive research efforts, none of the processes has been yet on stream due to two common drawbacks of catalysts (i) noble metal showed lower selectivity towards propanediol and (ii) Cu based catalysts although gave better selectivity to propanediol it has a health hazard due to Cr associated with it. Glycerol hydrogenolysis involves a series and complex reaction steps forming a complex reaction pathway, and also the kinetics of glycerol hydrogenolysis has not been well established as evidenced by a small number of publications on these topics [23, 24].

Among several industries, Senergy has licensed Suppes' technology and Archer Daniels Midland (ADM) announced a glycerol to 1,2-PDO production plant having a capacity of 100,000 tons per year and is closely linked to ADM's existing biodiesel production. Some other industries also have announced 1,2-PDO production plants and the details are summarized in Table 1.

TABLE 1 Summary of commercial plants announced for hydrogenolysis of glycerol to 1,2-PDO.

Company	1,2-PDO Capacity	References
Senergy Chemical	65 million lb/yr	[2]
Dow Chemical Co.	157 million gal/yr	[4]
Dow Haltermann Custom Processing (DHCP)	145 million lb/yr	[4]
Archer Daniels Midland(ADM)	100,000 tpa	[2]
Huntsman Corp.	145 million lb/yr	[4]
Ashland and Cargill	65,000 metric ton/yr	[4]
Virent Energy Systems, Inc.	Not mentioned	[4]

Considering all these facts, there is still a need to design more stable, highly active and selective, inexpensive, catalyst for hydrogenolysis of glycerol to 1,2-PDO. This chapter is mainly focused on our work on the development of heterogeneous chromium containing as well as non-chromium Cu catalysts for glycerol hydrogenolysis to 1,2-PDO. Copper chromite catalysts include those with different promoters *viz.* Ba, Al, Zn, and Mg for selective hydrogenolysis of glycerol to 1,2-PDO. All these catalysts were prepared, characterized and their preliminary activity results with respect to conversion and product distribution were obtained in laboratory scale batch and continuous fixed bed reactors. This study includes process optimization, raw material consumption and catalyst stability studies that will be significant for scale-up of the process.

2.3 EXPERIMENTAL SECTION

Non chromium copper catalyst developed in our laboratory was having Al as the only other component along with copper. Various CuAl compositions were prepared by simultaneous co-precipitation and digestion technique that involved desired concentrations of aqueous Cu $(NO_3)_2.3H_2O$, and Al $(NO_3)_3.9H_2O$ as metal precursors and using an aqueous potassium carbonate as a precipitating agent. The blue precipitate thus obtained was dried in static air oven at 373K for 8 hr followed by calcinations and activation. A detailed preparation procedure is given by Mane et al. [25].

Several copper chromite catalysts containing one or more promoters were prepared using ammonium chromate for precipitation from the nitrate precursors of the respective metals. The brown precipitate thus obtained was subjected to drying, calcinations and activation. A typical preparation procedure is reported elsewhere [26]. For the purpose of comparing the activity of our catalysts with that of Ba containing copper chromite as a bench mark catalyst, the same was also prepared by a patented procedure [27]. The prepared catalysts were characterised to ascertain the various phases of Cu formed after activation and the surface acidity by XRD and NH_3 TPD techniques respectively. X-ray powder diffraction patterns were recorded on a Rigaku, D-Max III VC model, using nickel filtered CuKα radiation. The samples were scanned in the 2θ range of 1.5–80°. The crystallite sizes of Cu species in various samples were determined using Scherrer–Warren equation given below.

$$< L >= \frac{K\lambda}{\beta \cos\theta}$$

where,
<L> = measure for the dimension of the particle in the direction perpendicular to the reflecting plane

λ = X-ray wavelength, nm
β = peak width
θ = angle between the beam and the normal on the reflecting plane, degree
K = constant that is (0.9)

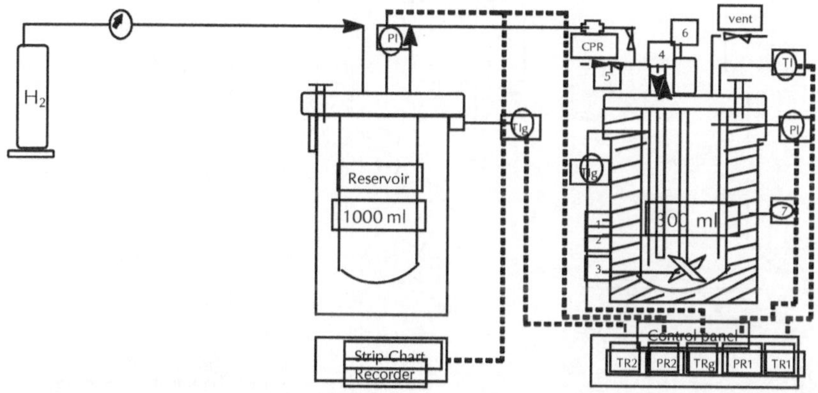

1. Reactor 2. Stirrer shaft 3. Impeller 4. Cooling water
5. Sampling valve 6. Magnetic stirrer 7. Electric furnace
TI: Thermocouple, PI: Pressure transducer, Tlg : Thermocouple for gas, N_2 : Nitrogen cylinder
PR: Pressure regulator, CPR: Content pressure regulator,
TR1: Reactor temperature indicator, PR1: Reactor pressure indicator, TR2: Reservoir temperature indicator
TRg : Gas temperature indicator PR2: Reservoir pressure indicator

FIGURE 4 Schematic of batch slurry reactor.

The TPD measurements were carried out on a Quantachrome CHEMBET 3000 instrument. In order to evaluate acidity of the catalysts, ammonia TPD measurements were carried out by:

(i) Pre-treating the samples from room temperature to 473K under nitrogen flow rate of 65 ml/min.
(ii) Adsorption of ammonia at room temperature.
(iii) Desorption of adsorbed ammonia with a heating rate of 283K min^{-1} starting from the adsorption temperature to 973K.

In order to demonstrate the efficiency of the prepared catalyst and typical product distribution in glycerol hydrogenolysis, the catalyst in powder form was used in high pressure batch slurry Parr reactor (300 ml), the schematic of which is shown in Figure 4. Typical hydrogenolysis conditions were temperature, 493K; glycerol concentration, 20 wt %; catalyst loading, 1 g; and hydrogen pressure 35–70 bar. The prepared catalysts were pre-reduced under H_2 at 473K for 12 hr.

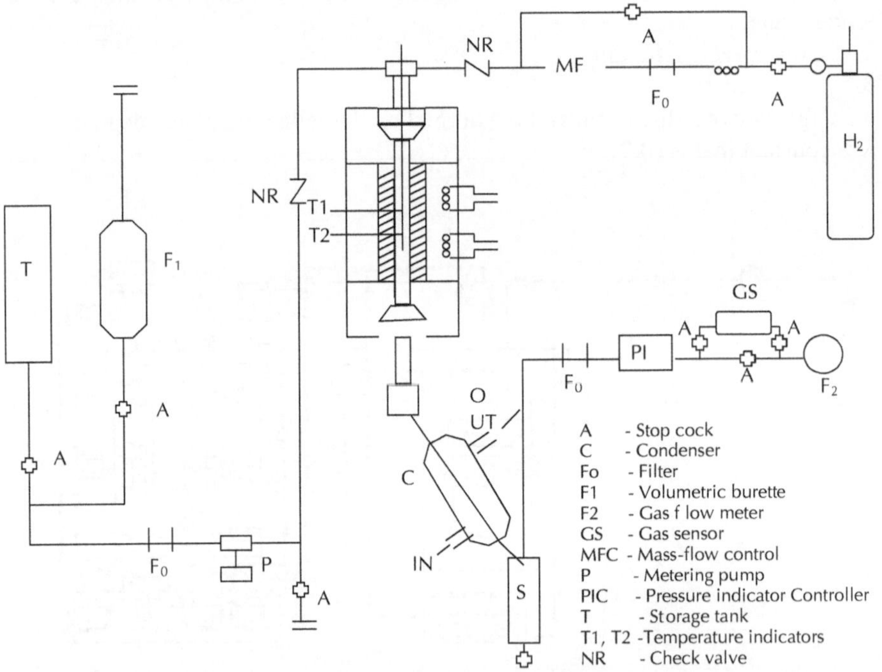

FIGURE 5 Schematic of high pressure fixed bed continuous reactor.

Continuous hydrogenolysis of glycerol was carried out in a bench scale, high pressure, fixed bed reactor supplied by M/s Geomechanique, France. A schematic of the reactor setup is shown in Figure 5. The powdered catalyst was pelletized in the form of pellets of 1×10^{-2} m diameter and cut into 4 pieces each having 2.5×10^{-3} m diameter 23 g of the pelletized catalyst was charged into the reactor. The catalyst bed of 0.11 m was packed in the middle of the reactor with carborundum as an inert packing above and below the catalyst bed. After flushing the reactor first with N_2 and then with H_2 at room temperature, the reactor was pressurized with H_2 at the desired temperature and the liquid feed was started to initiate the reaction. Liquid samples withdrawn at regular intervals of time were analyzed by GC (Varian 3600) equipped with a flame ionization detector and a capillary column (HP-FFAP 30 m, 0.53 mm, 1 μm).

2.4 DISCUSSION AND RESULTS

Since, 1,2-PDO formation *via* glycerol hydrogenolysis involves selective cleavage of C-O bond without breaking C-C bond, more focus has been given on developing copper based catalysts due to their high efficiency for C-O bond hydrogenolysis and poor activity for C-C bond cleavage [25]. Nevertheless, as shown in Scheme 3, first step in hydrogenolysis of glycerol to 1,2-PDO involves formation of either acetol or glyceraldehyde depending on acidic or basic sites of the catalysts respectively and reaction conditions. Between these two pathways, dehydration route to 1,2-PDO is more commonly followed due to simple preparation methods of catalyst systems having inherent acidic properties. In addition, acetol is also an important intermediate in hydrogen production by catalytic steam reforming [17, 28-30], pyruvaldehyde synthesis through oxidation [31] and as a starting material in various organic transformations [32, 33].

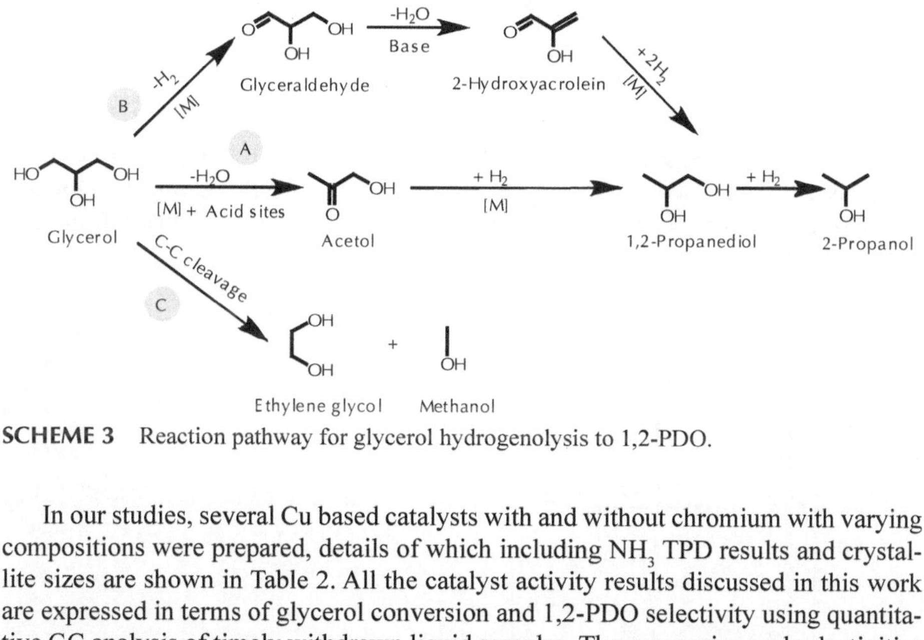

SCHEME 3 Reaction pathway for glycerol hydrogenolysis to 1,2-PDO.

In our studies, several Cu based catalysts with and without chromium with varying compositions were prepared, details of which including NH_3 TPD results and crystallite sizes are shown in Table 2. All the catalyst activity results discussed in this work are expressed in terms of glycerol conversion and 1,2-PDO selectivity using quantitative GC analysis of timely withdrawn liquid samples. The conversion and selectivities were calculated as follows:

$$\text{Conversion}\,(\%) = \frac{\text{initial glycerol concn.} - \text{final glycerol concn.}}{\text{initial glycerol concn.}} \times 100$$

$$\text{Selectivity}(\%) = \frac{\text{concn. of a product formed}}{\text{total concn. of all products}} \times 100$$

TABLE 2 Composition and physico-chemical properties of prepared catalysts. *[27]

Catalysts	Composition	Total NH_3 desorbed (mmole/g)	Crystallite size (nm)
Cu Al-1	50% Cu; 50% Al	1.567	7
Cu Al-2	30% Cu; 70% Al	1.553	-
Cu Al-3	70% Cu; 30% Al	1.342	156
CuCR-1	67% Cu; 33% Cr	0.3414	95
CuCr-2	54% Cu; 26%Cr; 20%Al	0.856	149
CuCr-3	54% Cu; 16%Cr; 30%Al	0.7856	69
CuCr-4	41% Cu; 29%Cr; 30%Ba	1.2444	-
CuCr-5	25% Cu; 25%Cr; 39% Zn; 11% Al	0.7935	-
CuCr-6	25% Cu; 25%Cr; 11% Zn; 39% Al	0.9563	-
CuCr-7	40% Cu; 30% Cr; 20% Al; 10% Ba	1.0235	-
CuCr-8*	60%Cu; 33%Cr; 7%Ba	1.184	-

A typical gas chromatogram of reaction crude of glycerol hydrogenolysis using our catalyst system is shown in Figure 6. As can be seen from this chromatogram, the only by-products formed were acetol and ethylene glycol (EG) and that too in very minor quantities.

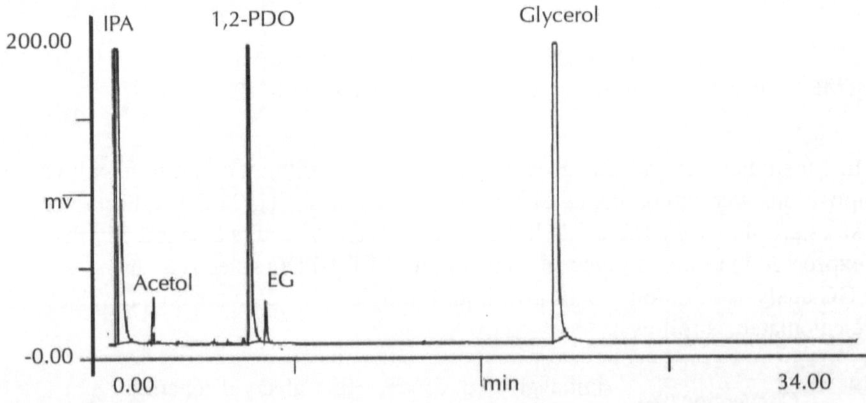

FIGURE 6 A typical gas chromatogram of glycerol hydrogenolysis crude.

The screening results of the prepared catalysts for glycerol hydrogenolysis in batch slurry reactor are shown in Figure 7.

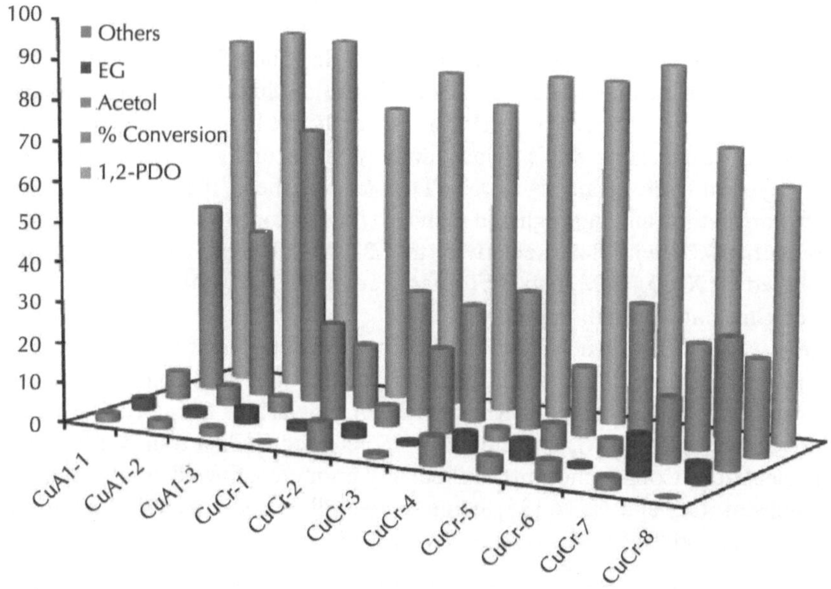

Figure 7. Catalyst screening for glycerol hydrogenolysis.
Reaction conditions: temperature, 493K; 20 wt% glycerol; pH2, 52 bar; solvent, isopropanol (IPA); catalyst, 0.01g/ml; reaction time, 5 hr.

As non promoted CuCr-1catalyst showed the lowest glycerol conversion of 16%, Al and Zn were incorporated as promoters into Cu-Cr catalyst with the expectation that their Lewis acid characteristic would enhance glycerol hydrogenolysis activity, which in fact was observed (CuCr-2 and 3, Figure 7). The higher activities of CuCr-2 and 3 catalysts than that of CuCr-1 could be attributed to their increased acidity (0.856 and 0.7856 mmol NH_3/g) over that of CuCr-1 (0.3414 NH_3/g). A very interesting observation was made when Al and Zn were together incorporated into CuCr-1, in which higher amount of Al (39%, CuCr-6) gave two times higher activity (34% conversion) than that of CuCr-5 (Al 11%). This confirms that not only acidity of the catalysts but also the appropriate Al concentration is necessary which might be affecting the glycerol hydrogenolysis kinetics. In case of Ba promoted CuCr catalysts (CuCr-4), highest glycerol conversion of 34% was achieved with > 84% selectivity to 1,2-PDO, which was in accordance with its highest acidity (1.2444 mmol NH_3/ g). On the other hand, addition of Al along with Ba (CuCr-7) although showed similar activity to that of CuCr-4, selectivity to 1,2-PDO was only 71% due to accumulation of acetol. Thus again minimum concentration of Ba was found to be necessary for forming hydrogenation product. The performance of our CuCr-4 catalyst was also found to be better than that of patented CuCrBa (CuCr-8) catalyst, especially with regard to 1,2-PDO selectivity. The difference in performance could be attributed to the use of carboxylic acid in the catalyst preparation procedure [27].

Our attempts to replace chromium containing catalysts led to the development of efficient CuAl catalyst systems with varying CuAl compositions. All these CuAl catalysts showed much higher activity than that of the best CuCr-4 and CuCr-6 catalysts due to pronounced acidity of the former (Table 2). In spite of similar acidity of CuAl catalysts, highest Cu content of 70% in CuAl-3 resulted into highest glycerol conversion of 69% and 1,2-PDO selectivity (90%). Recently, we have also developed several catalyst formulations using waste fly ash along with Cu, and was reported for the first time for glycerol hydrogenolysis to 1,2-PDO. Among these, the alkali pretreated fly ash by fusion method and impregnated with Cu showed considerable activity (glycerol conversion of 37%) with 84% selectivity to 1,2-PDO. The activity results were discussed based on XRD, TEM, and BET characterization [34]. Further work on fly ash supported Cu catalyst is still in progress.

As discussed, although acidity of the catalysts was important in glycerol hydrogenolysis reaction, equally important is the identification of various phases in multi component catalyst systems. Hence, XRD patterns of the various activated CuCr catalyst samples without and with Al, Ba, and Zn promoters and CuAl are shown in Figures 8 and 9 respectively. Copper chromite without any promoter showed predominant peaks at 2θ values of 43.2 and 50.1o [35] while two small peaks at 35.4 and 38.4o due to metallic copper and Cu+2 phases respectively [19].

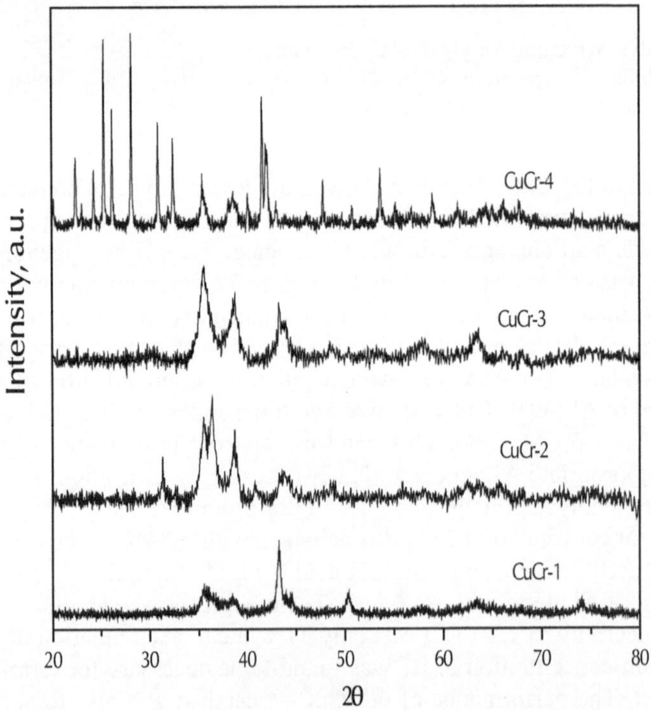

FIGURE 8 XRD patterns of CuCr catalyst with and without promoters.

Al promoted CuCr-2 showed a major peak at 2θ = 36.4° due to Cu^{+1} phase and two minor broad peaks at 2θ = 35.4 and 38.4° for Cu^{+2}. The CuCr-3 catalyst with Zn also showed the presence of Cu^{+2} with a small peak of metallic copper. While the XRD spectrum of the Ba promoted copper chromite catalyst showed peaks at 2θ values of 22.4, 25.4, 28.2, 41.6, 41.9, 43, and 30.9° which correspond to $BaCrO_4$ phase [36]. A peak at 2θ value of 43.2° was due to metallic copper while no peak corresponding to Cu^{+2} or Cu^{+1} phases was observed. This clearly showed that the presence of Ba was responsible for stabilizing Cu state in copper chromite catalysts which is mainly responsible for hydrogenolysis of glycerol as indicated by the activity results shown in Figure 7. It is also interesting to note that crystallite sizes obtained from diffraction peak for Cu (2θ = 43.2°) state in CuCr catalysts without and with promoters varied as follows: 156 nm (CuCr-1) > 149 nm (CuCr-3) > 95 nm (CuCr-2) > 69 nm (CuCr-4). Thus Ba was found to be responsible for inhibiting the agglomeration which also has been reported earlier [37].

The XRD patterns of CuAl-1 catalysts samples after calcinations (Figure 9, A1) showed dominant peaks at 2θ values of 35.46 and 38.7° corresponding to Cu^{2+} [19]. While the reduced CuAl-1catalyst shows broader peaks at 2θ =36.54 and 43.36° (Figure 9, A2) which were assigned to Cu^{+} and metallic Cu phases respectively, indicating the inadequate reduction of Cu. This also confirms that the Cu^{2+} in CuAl catalyst undergoes the sequential reduction as CuO first reduced to Cu_2O cubic phase as a stable intermediate and then to metallic Cu [35]. Using Scherrer–Warren equation, the crystallite sizes of CuAl-1 catalyst was found to be 7 nm (11 nm by HRTM) while those of bulk catalysts were in a range of 69–156 nm respectively. This shows that the extent of aggregation for CuAl-1 catalyst was much less than that observed for the bulk catalyst under reaction conditions which could be due to the presence of Cu^{+} species in CuAl-1 catalyst.

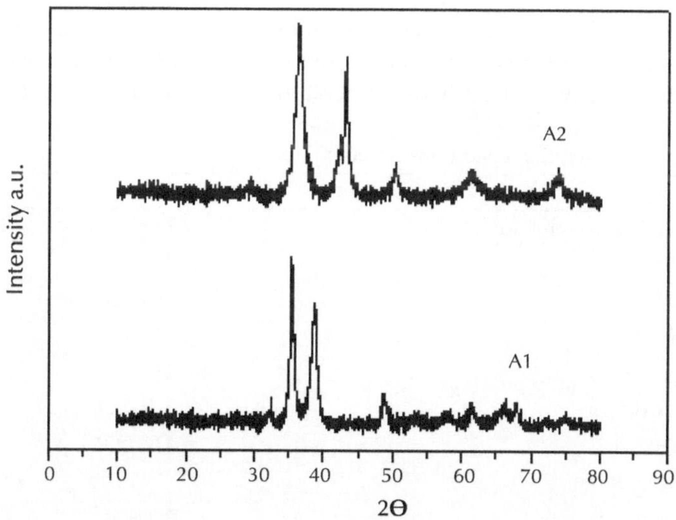

FIGURE 9 XRD pattern of CuAl-1 catalyst (A1) calcined, (A2) reduced.

It was also thought appropriate to screen the activity of best catalysts for hydrogenolysis of glycerol in water. For this purpose, we selected non chromium CuAl-1, and chromium containing with and without Ba promoter *viz.* CuCr-4 and CuCr-1 catalysts. These studies revealed that the activity of CuAl-1 catalyst was more than three times higher than those of any of the chromium containing catalysts for glycerol hydrogenolysis in water. Glycerol conversion obtained over CuAl-1 catalyst was 38% with 91% selectivity to 1,2-PDO while the chromium containing catalysts gave glycerol conversion in a range of 9–18% with dramatic decrease in 1,2 PDO selectivity to 38% due to the accumulation of an intermediate, acetol indicating slower kinetics of acetol hydrogenation in water. The activity and selectivity exhibited by CuAl-1 catalyst in water was almost comparable to that in IPA solvent indicated the better water tolerance of the CuAl-1 catalyst hence, aqueous glycerol solution can be directly used as a feed for its selective hydrogenolysis to 1,2-PDO.

As discussed, batch hydrogenolysis studies showed the highest activity and 1,2-PDO selectivity in IPA and water media for our CuCr-4 (Ba promoted) and non chromium CuAl-1 catalysts respectively. In addition, catalyst stability also needs to be established hence time on stream activity (TOS) of CuCr-4 (Ba promoted) was evaluated and compared with that of CuCr-1 (without promoter) for the continuous hydrogenolysis of glycerol. In order to study TOS first, the effects of temperature, H_2 pressure, liquid and gas flow rates and glycerol concentration on the conversion of glycerol and the product selectivities were also investigated under the range of reaction conditions as given in Table 3. From these parameter effect studies, optimum reaction conditions achieved for maximum conversion (65%) and 1,2-PDO selectivity (85-91%) are given in Table 4 which were then used for TOS studies. Figure 10(a) shows that CuCr-4 catalyst gave an excellent performance for 800 h as against abrupt drop in activity of CuCr-1 catalyst in <80h (Figure 10(b)) for continuous operation in IPA under 493 K, and 40 bar hydrogen pressure conditions. The highest activity of barium promoted CuCr-4 catalyst was mainly due to the following reasons. (i) Higher acidity responsible for catalyzing the first step of dehydration of glycerol to acetol, more efficiently. (ii) Formation of $BaCrO_4$ phase that inhibits the growth of crystallites contributing to prolonged catalyst activity under high temperature reaction conditions [38].

TABLE 3 Range of operating conditions for continuous.

Parameter	Range
Initial concentration of glycerol, wt%	20-60
Solvent	IPA
Temperature, K	453-513
H2 pressure, bar	20-60
Catalyst wt, g	23 g
Liquid velocity, LHSV, h^{-1}	0.78-2.34
Gas velocity, GHSV, h^{-1}	434-1304
Catalyst packing length, m	0.11
Particle diameter, dp, m	0.004
Density of the catalyst, kg/m^3	1685

TABLE 4 Optimum reaction parameters hydrogenolysis of glycerol to 1,2-PDO.

Parameters	Optimum conditions
Temperature	493K
H2pressure	4.2 MPa
Feed flow rate	30 mL/h
Gas flow rate	10 NL/h
Glycerol loading	20 wt%

Similarly TOS activity of CuAl-1 catalyst studied in water showed the consistent performance for 400h with glycerol conversion, > 72% and 1,2-PDO selectivity of 90%.

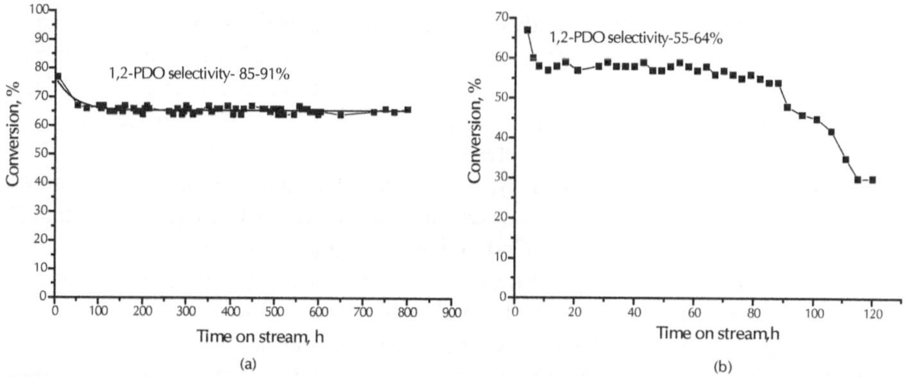

FIGURE 10 Time on stream activity of (a) CuCr-1 and (b) CuCr-4 for hydrogenolysis of glycerol to 1,2-PDO.

Reaction conditions: catalyst wt., 23 g; solvent, IPA; 20 wt % glycerol; feed flow rate, 30 ml/hr; H_2 flow rate, 10 nl/hr; pressure, 40 bar; temperature, 493K.

The CuCr-4 catalyst performance was also compared with the reported literature (Table 5) which revealed that CuCr-4 catalyst showed a better performance. Hence, considering very conservative values of conversion, selectivity and catalyst life, the raw material consumption for producing 1 kg of 1,2-PDO was also evaluated for our catalyst from process point of view, which is shown in Table 6.

TABLE 5 Comparison of CuCr-4 catalyst with literature data [39].

Catalyst	Substrate	Added base	Temp K	Pressure (bar)	t/t Feed PG
Ni+Re/C+NaOH	Glycerol	NaOH	503	89.6	0.55
7% Ni+1%Re/C	Glycerol	NaOH	503	89.6	0.65
Homogeneous Nobel Metal	Glycerol	–	523	68.9	0.717
CuCr [40]	Glycerol	–	503	13.7	0.56
NCL catalyst	Glycerol	–	493	51.7	0.60
NCL catalyst	Glycerol	–	493	41.3	0.75*

*Our results

TABLE 6 Raw material consumption based on CuCr-4 for glycerol to 1,2-PDO.

Raw material	Requirement kg for per kg 1,2-PDO
Refined Glycerol	1.67
Hydrogen	0.036
Catalyst	0.008

2.5 CONCLUSION

This chapter summarizes our work on catalyst development for reuse of glycerol by-product to a value added commodity chemical such as 1,2-PDO. The promoters *viz.* Al, Zn, and Ba in copper chromite catalysts extensively affect the activity as well as the selectivity in hydrogenolysis of glycerol. NH_3 TPD characterization showed that the higher acidity of Al and Zn promoted CuCr catalysts (0.856 and 0.7856 mmole NH_3/g) than that of CuCr-1 (0.3414 NH_3/g) would be responsible for their higher activities. Also CuCr catalyst with Ba as a promoter showed the highest conversion and 1,2-PDO selectivity of 34 and 85% respectively, in a batch operation. While, in a continuous operation it gave further increase in glycerol conversion to 65% with a higher stability (TOS, 800 hr^{-1}) as compared to CuCr-1. Highest activity, selectivity and stability of Ba promoted CuCr catalysts was mainly due to its enhanced acidity catalyzing the first step of dehydration of glycerol to acetol and stabilization of the crystallite size of Cu at a lower value of 69 nm caused by $BaCrO_4$ phase formation. A non-chromium copper catalyst (CuAl) was also developed for aqueous glycerol hydrogenolysis giving 69% conversion with 90% selectivity to 1,2-PDO. The higher activity and stability of CuAl-1 catalyst was explained by the presence of Cu^+ which inhibited the sintering of the active phase (Cu) under reaction conditions.

KEYWORDS

- **Byproducts**
- **Copper Chromite Catalysts**
- **Gallons**
- **Hydrogenolysis**
- **Scherrer–Warren Equation**
- **Transesterification**

REFERENCES

1. Nakagawa, Y. and Tomishige, K. Heterogeneous catalysis of the glycerol hydrogenolysis. *Catal. Sci. Technol.*, **1**, 179–190 (2011).
2. Bozell, J. J. and Petersen, G. R. Technology development for the production of biobased products from biorefinery carbohydrates—the US Department of Energy's "Top 10" revisited. *Green Chem.*, **12**, 539–554 (2010).

3. Bozga, E. R., Plesu, V., Bozga, G., Bildea, C. S., and Zaharia, E. Conversion of glycerol tp propanediol and acrolein by heterogeneous catalysis. *Rev. Chim.*, **62**(6), 646–654 (2011).
4. www.aiche.org/cep
5. www.abginc.com
6. Hass, T., Neher, A., Arntz, D., Klenk, D., and Girke, W. Method of preparation of 1,2- and 1,3-propanediol. EP0598228 A1 (1994).
7. Dasari, M. A., Goff, M. J., and Suppes, G. J. Noncatalytic alcoholysis kinetics of soyabean oil. *J. Am. Oil Chem. Soc.*, **80**, 189–192 (2003).
8. Zhou, C. H., Beltramini, J. N., Fana, Y. X., and Lu, G. Q. Chemoselective catalytic conversion of glycerol as a biorenewable source to valuable commodity chemicals. *J. Chem. Soc. Rev.*, **37**(3), 527–549 (2008).
9. Furikado, I., Miyazawa, T., Koso, S., Shimao, A., Kunimori, K., and Tomishige, K. Catalytic performance of Rh/SiO_2 in glycerol reaction under hydrogen. *Green Chem.*, **9**, 582–588 (2007).
10. Feng, J., Fu, H., Wang, J., Li, R., and Chen, H. Hydrogenolysis of glycerol to glycols over ruthenium catalysts: Effect of support and catalyst reduction temperature. *Catal. Commun.*, **9**, 1458–1464 (2008).
11. Miyazawa, T., Koso, S., Kunimori, K., and Tomishige, K. Development of a Ru/C catalyst for glycerol hydrogenolysis in combination with an ion-exchange resin. *Appl. Catal. A*, **318**, 244–251 (2007a).
12. Miyazawa, T., Koso, S., Kunimori, K., and Tomishige, K. Glycerol hydrogenolysis to 1,2-propanediol catalyzed by a heat-resistant ion-exchange resin combined with Ru/C. *Appl. Catal. A*, **329**, 30–35 (2007b).
13. Alhanash, A., Kozhevnikova, E. F., and Kozhevnikov, I. V. Hydrogenolysis of glycerol to propanediol over Ru: Polyoxometalate bifunctional catalyst. *Catal. Lett.*, **120**(3–4), 307–311 (2008).
14. Balaraju, M., Rekha, V., Sai Prasad, P. S., Prabhavathi Devi, B. L. A., Prasad, R. B. N., and Lingaiah N. Influence of solid acids as co-catalysts on glycerol hydrogenolysis to propylene glycol over Ru/C catalysts. *Appl. Catal. A*, **354**(1–2), 82–87 (2009).
15. Ma, L., He, D., and Li, Z. Promoting effect of rhenium on catalytic performance of Ru catalysts in hydrogenolysis of glycerol to propanediol. *Catal. Commun.*, **9**(15), 2489–2495 (2008).
16. Chaminand, J., Djakovitch, L., Gallezot, P., Marion, P., Pinel, C., and Rosier, C. Glycerol hydrogenolysis on heterogeneous catalysts. *Green Chem.*, **6**, 359–361 (2004).
17. Dasari, M. A., Kiatsimkul, P. P., Sutterlin, W. R., and Suppes, G. J. Low-pressure hydrogenolysis of glycerol to propylene glycol. *Appl. Catal. A*, **281**(1–2), 225–231 (2005).
18. Perosa, A. and Tundo, P. Selective hydrogenolysis of glycerol with raney nickel. *Ind. Eng. Chem. Res.*, **44**(23), 8535–8537 (2005).
19. Wang, S. and Liu, H. C. Selective hydrogenolysis of glycerol to propylene glycol on Cu–ZnO catalysts. *Catal. Lett.*, **117**, 62–67 (2007).
20. Yin, A., Guo, X., Dai, W., and Fan, K. The synthesis of propylene glycol and ethylene glycol from glycerol using Raney Ni as a versatile catalyst. *Green Chem.*, **11**, 1514–1516 (2009).
21. Zhou, J., Guo, L., Guo, X., Mao, J., and Zhang, S. Selective hydrogenolysis of glycerol to propanediols on supported Cu-containing bimetallic catalysts. *Green Chem.*, **12**, 1835–1843 (2010).
22. Ryneveld, E., Mahomed, A. S., Heerden, P. S., Green, M. J., and Friedrich, H. B. A catalytic route to lower alcohols from glycerol using Ni-supported catalysts. *Green Chem.*, **13**, 1819–1827 (2011).
23. Xi, Y., Holladay, J. E., Frye, J. G., Oberg, A. A., Jackson, J. E., and Miller, D. J. A kinetic and mass transfer model for glycerol hydrogenolysis in a trickle-bed reactor. *Org. Proc. Res. and Develop.*, **14**(6), 1304–1312 (2010).
24. Torres, A., Roy, D., Subramaniam, B., and Chaudhari, R. V. Kinetic modeling of aqueous-phase glycerol hydrogenolysis in a batch slurry reactor. *Ind. Eng. Chem. Res.*, **49**(21), 10826–10835 (2010).
25. Mane, R. B., Hengne, A. M., Ghalwadkar, A. A., Vijayanand, S., Mohite, P. H., Potdar, H. S., and Rode, C. V. Cu:Al Nano Catalyst for Selective Hydrogenolysis of Glycerol to 1,2-Propanediol. *Cat. Lett.*, **135** (1–2), 141–147 (2010).

26. Rode, C. V., Ghalwadkar, A. A., Mane, R. B., Hengne, A. M., Jadkar, S. T., and Biradar N. S. Selective Hydrogenolysis of glycerol to 1,2-Propanediol: comparison of batch and continuous process operations. *Org. Process Res. Dev.*, **14**(6), 1385–1392 (2010).
27. Henkelmann, J., Becker, M., Bürkle, J., Wahl, P., Theis, G., and Maurer, S. Process for the preparation of 1,2-propanediol. WO 2007099161 A1 (2007).
28. Chiu, C. W., Dasari, M. A., Suppes, G. J., and Sutterlin, W. R. Dehydration of glycerol to acetol *via* catalytic reactive distillation. *AIChE J.*, **52**, 3543–3548 (2006).
29. Ramos, M. C., Navascue´s, A. I., Garcı´a, L., and Bilbao, R. Hydrogen Production by Catalytic Steam Reforming of Acetol, a Model Compound of Bio-oil. *Ind. Eng. Chem. Res.*, **46**(8), 2399–2406 (2007).
30. Yamaguchi, A., Hiyoshi, N., Sato, O., Rode, C. V., and Shirai, M. Enhancement of glycerol conversion to acetol in high-temperature liquid water by high-pressure carbon dioxide. *Chem. Lett.*, **37**, 926–927 (2008).
31. Ai, M. and Ohdan, K. Formation of pyruvaldehyde (2-Oxopropanal) by oxidative dehydrogenation of hydroxyacetone. *Bull. Chem Soc. Jpn.*, **72**(9), 2143–2148 (1999).
32. Paradowska, J., Rogozinska, M., and Mlynarski, J. Direct asymmetric aldol reaction of hydroxyacetone promoted by chiral tertiary amines. *Tetra Lett.* **50**(14), 1639–1641 (2009).
33. Wu, X., Ma, Z., Ye, Z., Qian, S., and Zhaob, G. Highly efficient organocatalyzed direct asymmetric aldol reactions of hydroxyacetone and aldehydes. *Adv. Synth. Catal.*, **351**, 158–162 (2009).
34. Rode, C. V., Mane, R. B., Potdar, A. S., Patil, P. B., Niphadkar, P. S., and Joshi, P. N. Copper modified waste fly ash as a promising catalyst for glycerol hydrogenolysis. Submitted (2011).
35. Pike, J., Chan, S., Zang, F., Wang, X., and Hanson, J. Formation of stable Cu_2O from reduction of CuO nanoparticles. *Appl. Catal. A*, **303**(2), 273–277 (2006).
36. Yan, Y., Wu, Q. S., Li, L., and Ding, Y. P. Simultaneous synthesis of dendritic superstructural and fractal crystals of $BaCrO_4$ by vegetal bi-templates. *Crystal Growth and Design*, **6**(3), 769–773 (2006).
37. Choudhary, V. R. and Pataskar, S. G. Thermal decomposition of ammonium copper chromate: effect of the addition of barium. *Thermochimica. Acta.*, **95**(1), 87–98 (1985).
38. Mane, R. B., Ghalwadkar, A. A., Hengne, A. M., Suryawanshi, Y. R., and Rode, C. V. Role of promoters in copper chromite catalysts for hydrogenolysis of glycerol. *Catal. Today*, **164**, 447–451 (2011).
39. Crabtree, S. P., Lawrence, R. C., Tuck, M. W., and Tyers, D. V. Optimize glycol production from biomass. *Hydrocarbon Process.*, 87–92 (February, 2006).
40. Suppes, G. J., Sutterlin, W. R., and Dasari, M. Method of producing lower alcohols from glycerol. US0244312 A1 (2005).

3 Recycling of Gray Wastewater from Household Laundering

Joydeep Paul Majumder and Dr. Anjali Pal

CONTENTS

3.1 INTRODUCTION

A significant amount of gray wastewater is generated in households from washing and laundering. This wastewater contains anionic surfactants (AS) and detergents as the main polluting agent along with some turbidity. Due to low organic content, this wastewater do not turn septic and hence can find some other use if the AS and the turbidity are removed. The AS present in wastewater cause major problem due to foaming. It also gets deposited within tissues of living organisms causing severe metabolic disorders Turbidity imparts unpleasantness to water and harbors a site for breeding of pathogenic microorganisms.

The AS are surface active compounds which can be removed by adsorption. Quite a large number of natural and artificial adsorbents are used for AS removal. Burnt clay has been found very useful for the removal of surfactants in both batch and continuous (column) studies. Crushed to the size of 500 microns, they have achieved removal efficiency of AS up to 84% in the alkaline pH range of pH 911 in batch studies conducted with real waste water samples having AS concentrations in the range of 90100 mg/l. The kinetic studies have shown a second order kinetics with saturation time at 8 hr.

In column studies surfactant removal was achieved at 100% for a period of 72 hr at a flow rate of 10 ml/min. With the break point time of 80 hr treating a volume of 48 l of waste water in a column bed 17 mm diameter and 20 cm depth. The exhaust point in this case was at 90 hr treating 54 l of wastewater. Turbidity removal was achieved to 86% while 72% of the total solids were removed after treatment.

The treated wastewater can be used for gardening, house washing, and washing other areas where potable grade water is not required. In farm houses, it can also be used for irrigation of vegetables and fruit plants.

The vast amount of gray wastewater is generated in India due to household washing and laundering. More than 80% of the water used can be recycled for applications which do not require water of potable quality including watering of plants, washing of household premises and flushing of toilets. Most of the wastewater contains detergents and surfactants along with low concentration of organic matter and turbidity which is caused by the suspended particles present after washing. Kitchen wastewater contains appreciable amounts of dissolved organics and hence it needs considerable treatment; but laundering wastewater can be rendered useful if the surfactants and turbidity can be removed. Turbidity can be removed by simple filtering in a domestic scale where very high efficiency is neither affordable due to financial reasons nor needed. Surfactants can be removed by adsorption with natural or artificial materials which can be used in bulk requiring neither very sophisticated technology nor huge financial investment.

Natural substances like montmorillonite saturated with Ca^{2+} (which got bonded with anionic surfactant dodecyl benzene sulfonate) [1] and sepiolite (Mg^{2+} ions released from sepiolite leading to precipitation of the surfactant) [2] are good adsorbents which are very effective in surfactant removal. Sand sorption process in surfactant removal from ground water gave an efficiency of 70.11% as seen from the studies of Khan and Zareen [3]. Granular charcoal achieved removal capacities of upto 98% when used in treating cationic surfactants present in low strengths in dilute streams of 5×10^{-5} M [4]. Alumina is a good adsorbent with an efficiency of 94% for removing anionic surfactant from wastewater at very high concentrations of 8,068 ppm with optimum adsorbent dosage of 120 g/l and contact time of 1 hr. Various types of synthetic resins has high anionic adsorption capacity of 11.6 g surfactant/g of adsorbent at rinsing water concentration of 0.10.3 g of surfactant/kg of wastewater [5].

3.2 EXPERIMENTAL

Brick granule was the adsorbent as well as the filter material selected for the treatment of the laundering wastewater. Both batch and column studies were performed on field samples having anionic surfactant concentration within the range of 80–96 mg/l taken from both household as well as community washing and laundering.

Removal of turbidity was studied by making a settling bed and a filter bed of grain size smaller than 45 microns prior to adsorption and applying a head of 1 m.

The treatment followed here was a three step process in which the wastewater was first allowed to settle for 24 hr due to which most of the settleable solids (>80%)

settled and some surfactants (about 20–25%) which were bonded with the dirt particles were also removed.

The second step was passing the wastewater through the filter media in down flow mode removing nearly all the suspended solids (almost 100%) and a vast majority of colloidal solids.

The third and final step was the treatment in the adsorption column where most of the surfactants were removed while flowing in up flow mode along with the other ions.

3.3 RESULTS

The brick granules are inert earth material consisting mainly of silica (85%) with ferric oxide (11%) and traces of oxides of titanium and potassium. The surface morphology of brick is shown in Figure 1.

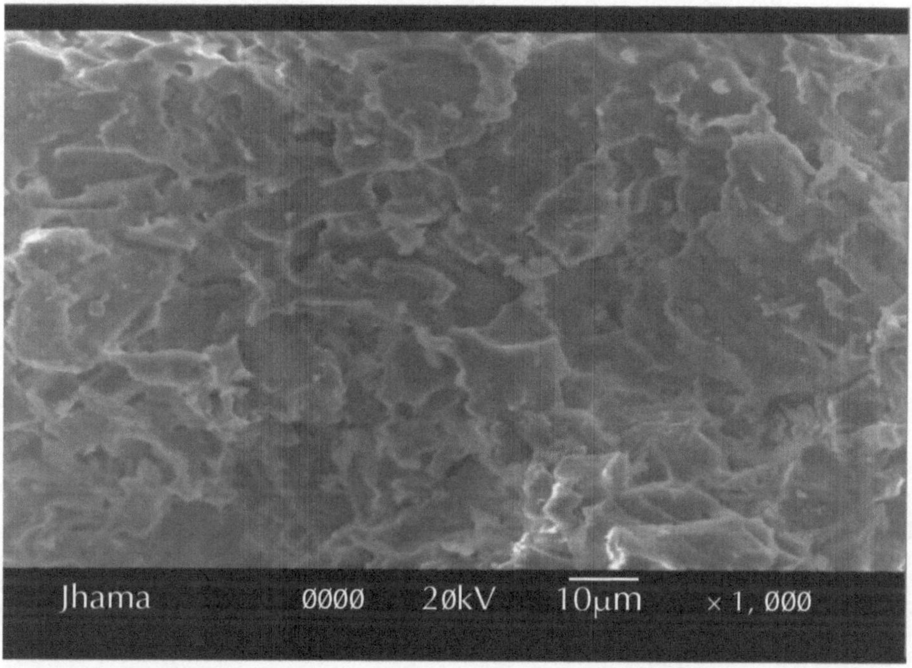

FIGURE 1 Surface morphology of brick granules at a magnification of 1,000 times.

The pH_{zpc} was found to be 6.35. The equilibrium time was 8 hr (Figure 2). The adsorption followed a second order reaction and hence concluded that it was a chemical adsorption.

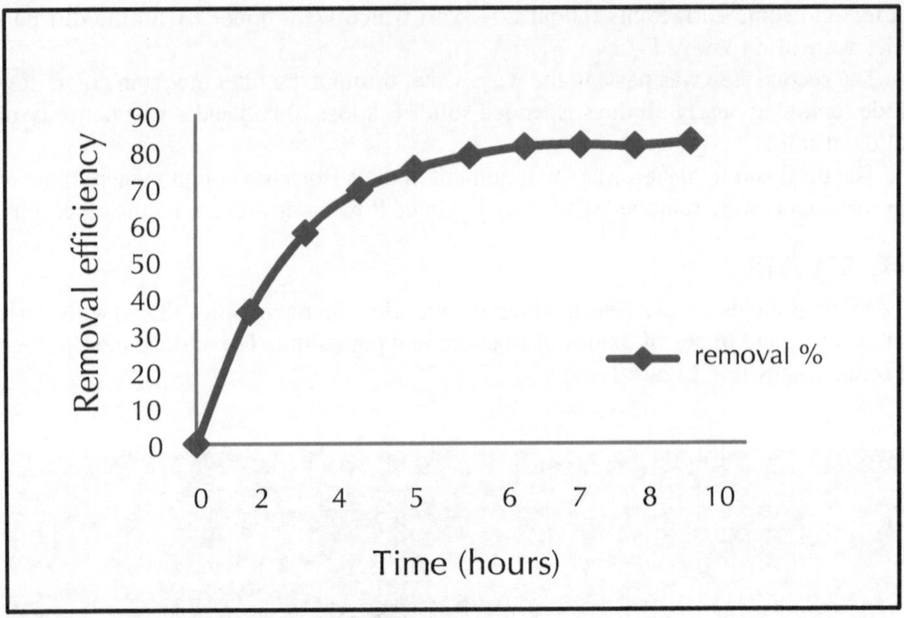

FIGURE 2 Equilibrium kinetics of brick granules.

The Langmuir isotherm was followed by brick granules as shown in Figure 3. The Q_{max} values were found to be 142.85 mg/g.

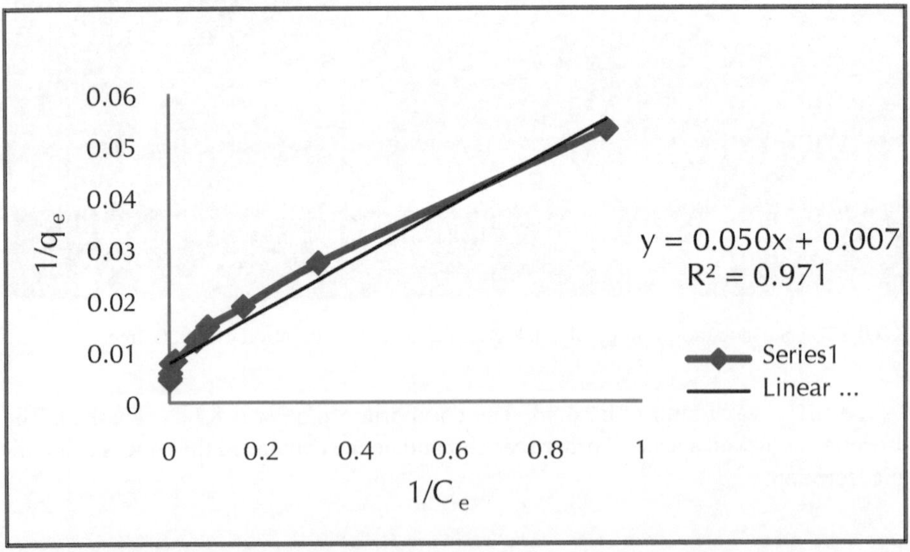

FIGURE 3 Langmuir isotherm of brick granules.

The column studies were conducted with column diameter of 1.6 cm and bed depth of 20 cm. it exhausted after 90 hr treating 541 of wastewater. The best efficiency (upto 80–84%) for surfactant removal by brick granules was seen at alkaline pH range from 8 to 11. The breakthrough curve of brick granules is shown in Figure 4.

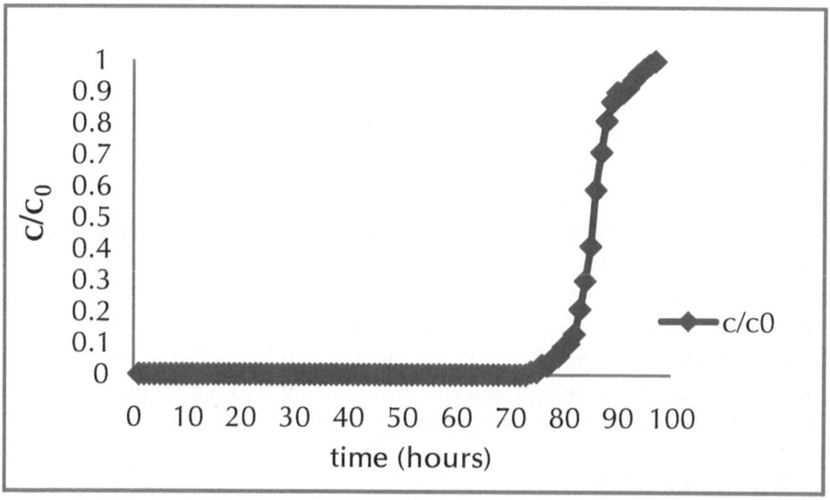

FIGURE 4 Breakthrough curve of brick granules.

The treatment of field samples having surfactant concentration of 90 mg/l along with other impurities is shown in Table 1. No leaching was observed from brick granules at all the pH values.

3.3.1 Recycling in Domestic Household

The small column filter units made of metal and filled with crushed brick sieved to obtain a size of 0.51 mm can be used in household kitchens and wash basins after passing through which the wastewater can be recycled in the flushes of toilets and water closets. The column should have the top layer filled with brick dust of size smaller than 100 microns so that efficient filtration is ensured to remove majority of the turbidity.

TABLE 1 Treatment efficiencies of various parameters in the wastewater with brick granules uptill the exhaustion point.

Parameter	Efficiency (%)
Turbidity	85.94
Anionic surfactant	94.09
Suspended solids	100
Dissolved solids	67.5
B. O. D.	90
C. O. D.	81.9

TABLE 1 *(Continued)*

Parameter	Efficiency (%)
Chloride	22.6
Sulphide	42
Sulphate	64.5
Nitrate	5.5
Odor	Significant

Alternatively a small pit of approximate size of 20 × 20 cm and suitable depth to retain the wastewater generated everyday can be dug in the ground and filled with brick granules of above mentioned size where the wastewater after washing of the clothes can flow in and undergo adsorption to remove the surfactants after which it can be useful as gardening water to plants. If up flow mode of treatment is used, it further eliminates the requirement of filter bed for turbidity removal as most of the settleable solids will settle at the bottom during the retention time of 1 day.

The waste water after the above treatment step can be stored in an underground reservoir from which it can be pumped and also used for washing of outside premises, roads, public parks (in community treatment), and other rough uses.

3.3.2 Disposal of Sludge

The brick granules can absorb surfactants nearly upto 1/10th of its weight. Knowing the strength of surfactant in wastewater, the time required for exhaustion of brick granules can be estimated and the exhausted material can be used in places where sand is used as the material has physical properties similar to sand. It can be used for land filling, mortar making with cement and as a sub base material for road construction.

3.4 CONCLUSION

Recycling of laundering wastewater will help to mitigate the problems of water shortages in areas where fresh water is not easily available. The wastewater after treatment can be safely used for various household and community purposes. The treatment systems can be designed both at household level as well as at community level. The exhausted adsorbent can also undergo different types of uses. A great advantage of this system is that it can be operated by people without very high technical knowledge and skills. The material being easily available poses no financial burden to its users.

KEYWORDS

- **Anionic Surfactants**
- **Brick Granule**
- **Burnt Clay**
- **Natural Substances**
- **Turbidity**
- **Wastewater**

REFERENCES

1. Yang, K., Zhu, L., and Xing, B. "Sorption of sodium dodecyl benzene sulphonate by montmorillonite." *Environmental Pollution*, **145**(2), 571576 Elsevier (January, 2007).
2. Ozdemir, O., Cinar, M., Sabah, E., Arslan, F., and Celik, M. S. "Adsorption of anionic surfactants onto sepiolite." *Journal of Hazardous Materials*, **147**(12), 625632, Elsevier (August, 2007).
3. Khan, M. N. and Zareen, U. "Sand sorption process for the removal of sodium dodecyl sulphate (anionic surfactant) from water." *Journal of Hazardous Materials*, **133**(13), 269275, Elsevier (May, 2006).
4. Saleh, M. M. "On the removal of cationic surfactants from dilute streams by granular charcoal." *Water Research*, **40**(5), 10521060, Elsevier (March, 2006).
5. Schouten, N., van der Ham, L. G. J., Euverink, G. J. W., and de Haan, A. B. "Selection and evaluation of adsorbents for the removal of anionic surfactants from laundry rinsing water." *Water Research*, **41**(18), 42334241, Elsevier (October, 2007).

4 Calcined Clay Applied in Concrete

Edisley Martins Cabral, Raimundo Pereira de Vasconcelos, Raimundo Kennedy Vieira, Nilton Souza. Campelo, Adalena Kennedy Vieira, Claudia. Candido. Silva, and Mirian Dayse Lima

CONTENTS

4.1 INTRODUCTION

The state of Amazonas is located in a large sedimentary basin with few places where there are outcrops of rock. Thus, the main cities in the region have difficulty in obtaining stoney material for use as coarse aggregate in construction.

The main source of such material is gravel which is obtained from river beds. However, cost of the material is very high because of the long distance that must be traveled to obtain it, which increases construction costs. Moreover, the extraction of gravel from river beds has caused siltation of rivers, a negative environmental impact. In order to find a more economically viable and environmental friendly material for coarse aggregate in concrete, it is worthwhile to study the viability of synthetic aggregates such as, calcined clay.

The term synthetic aggregate or Calcined Clay, refers to a material obtained from the processing of a soil or clayey material with satisfactory mechanical strength for a particular purpose. These characteristics are usually obtained by heating the ceramic body at high temperatures, above 760°C. The quality depends crucially on the ceramic raw material, firing temperature, and process of ceramic mass conformation [1]. In some cases, certain properties of ceramic products can be improved by using soils with a higher percentage of flux elements.

Calcined clay has been used to replace both fine and coarse aggregates as well as cement in concrete [2-4]. The use of crushed brick is a practice has been known since the beginning of civilization. In the case of coarse aggregate [5], the calcined clay has been obtained as waste from different industrial processes. The studies cited have addressed both environmental and economic questions related to the use of calcined clay.

Consider, the economic aspect, it must be noted that the cost of the aggregate is lower than that of the cement, so it is important to produce concrete with the greatest possible amount of aggregate and thus to reduce the consumption of cement. This assumption is valid only when the shipping cost is not high [6]. In this case, the use of calcined clay will be opportune because extensive areas of alluvial deposits, the main sources of raw material for the production of bricks and tiles, are located in the largest area of concentration of the red ceramic industry of the state of Amazonas (Brazil), which is near the biggest market for construction materials, the city of Manaus (capital of Amazonas state).

But cost is not the only reason for using this kind of aggregate. In addition to providing an alternative source of aggregate, the use of synthetic aggregate calcined clay (SACC) can reduce the environmental impact that is caused by extraction of natural aggregate in Amazonas.

Another important consideration in the use of SACC is the process by which it is produced. In the studies cited above, which used waste calcined clay, the process to produce the clay was dependent on the original industrial process. However, the calcined clay produced through this process was not designed to possess specific properties necessary for use as coarse aggregate. The aim of this study was to develop a controlled production process to produce SACC produced with Amazon soil to be applied in concrete.

4.2 MATERIALS AND METHODS

4.2.1 Method

The method used in this study is shown in overview form in Figure 1 and will describe in the text in sequence.

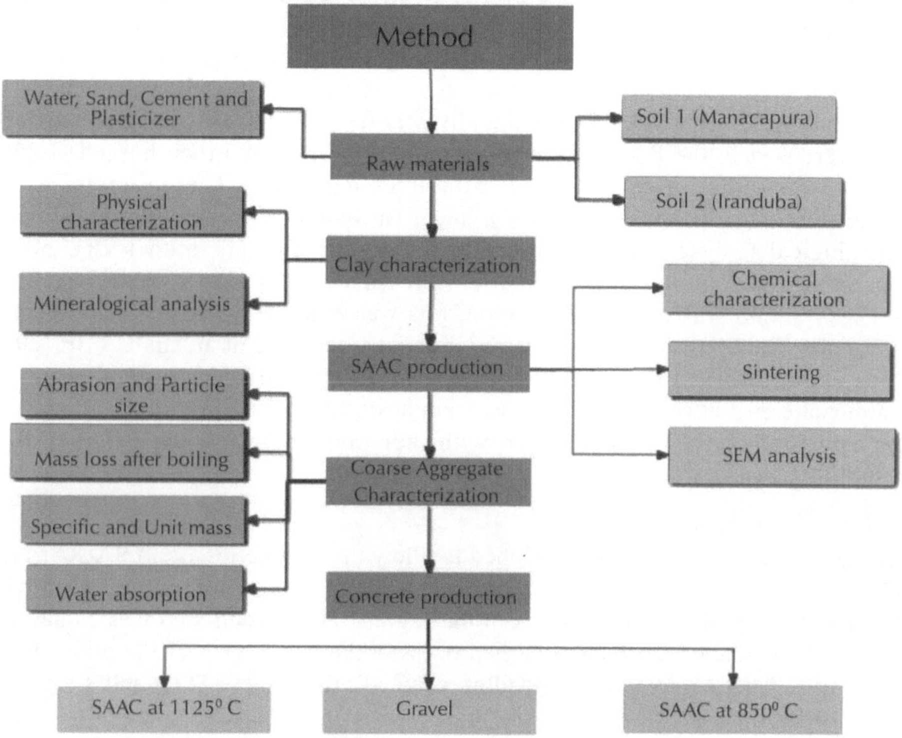

FIGURE 1 Methodological structure.

4.2.1.1 Raw Material

Clay samples were collected in two cities of the state of Amazonas, Manacapuru, and Iranduba, in deposits used for brick making. These clays were identified as Soil 1, collected in Iranduba city, and Soil 2, collected in Manacapuru city. Geological studies and preliminary tests carried out by the Society of Research and Mineral Resources in this region indicate the potential to produce large amounts of red ceramics. These studies estimate geological reserves of about 4.32 billion cubic meters [7].

The SACC was used in place of natural aggregate in concrete which also contained Portland cement CP32-Z, manufactured in Manaus, and natural river sand. The sand used in this study had a fineness modulus of 2.2 and was thus considered fine sand. The maximum diameter of this aggregate was 2.4 mm. The plasticizer used was a multifunctional additive, Tec-mult 400 of Rheotec. This additive meets the requirements of ASTM [8]. The water used in mixes was from the Amazon River, and presented a pH of 5.8.

4.2.1.2 Clay Characterization

Clay samples were characterized to determine their chemical, physical, and mineralogical properties.

The chemical composition of the soil sample used for the production of synthetic aggregate was determined by energy dispersive X-ray fluorescence with Shimadzu model EDX700HS. For the physical tests of the soil, prismatic specimens were cast in the dimensions 600 x 200 x 100 mm, and then burned at 850 and 1,125°C with 1 hr of soak time at the top temperature. These samples were then tested for water absorption (WA), linear shrinkage, and specific gravity. The mineralogical analyses were performed by X-ray diffractometer model XRD-6000 from Shimadzu. Samples of raw material were separated by size in centrifuge, and the clay fraction was collected. The analysis was made for the total sample and for the clay fraction. In samples where reflections occurred at an angle $<10°$ (2θ), which is indicative of the presence of expansive clay, the slide was subjected to an atmosphere of ethylene glycol in vacuum desiccators for a period of 12 hr, and then examined in the diffractometer, with radiation Cu $k\alpha$, which identified the crystalline phases.

4.2.1.3 SACC Production

Methods and equipment were developed to allow the production of an SACC from wet clay. As a starting point, a visit was made to potteries in order to acquire knowledge about the production of bricks in the region. The second step was to devise equipment capable of reproducing this process in the laboratory for the production of ceramics. This equipment included an adapted mill (Figure 2(a)) and a regular oven.

In the laboratory, samples of clay were put to dry in the shade to remove the excess moisture and soon afterwards they were harrowed manually with a mortar. Through this process the soil moisture was reduced to 28%. This value, as the limit of plasticity, tends to make the soil plastic and facilitates molding. Then the ceramic body was subjected to extrusion in an adapted mill, where the output of the conformed material was placed in rectangular nozzles with dimensions ranging from 4.8 to 12.5 mm. These upper and lower dimensions correspond to the maximum size found in the gravel that has been used as aggregate and to the restrictions of production lines, as can be seen in Figure 2(b).

It must be noted that the kilns used by the ceramic industry in Iranduba and Manacapuru are able to burn the raw materials used for production of bricks at 850 and 950°C. Thus, seeking a future use for this type of furnace for the production of SACC, the SACC was produced by burning at 850 and at 1,125°C. The latter is the temperature where the ceramic mass reached the lowest value for absorption of water and the maximum value for linear shrinkage to Soil 1. Morphological analysis of SACC processed at 850 and 1125°C was performed by scanning electron microscopy.

(a)

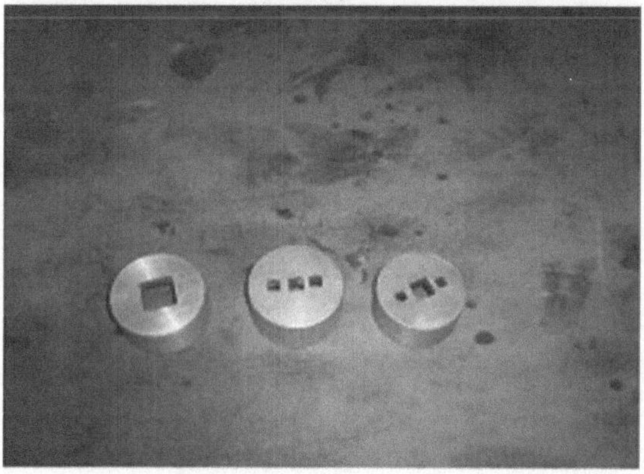

(b)

FIGURE 2. (a) Adapted mill for making clay aggregate. (b) Die mouthpiece for molding of clay aggregates.

4.2.1.4 Coarse Aggregate Characterization

The physical properties of the coarse aggregate were characterized to determine abrasion, particle size, mass loss after boiling, specific gravity, unit mass, and WA.

4.2.1.5 Concrete Production

Characteristics of concrete using gravel or SACC as coarse aggregate were studied. Analyses were done of characteristics such as, workability and compressive strength. For these analyses, several different mixes were made, varying the proportion of cement and aggregates. For each mix, the workability was verified (i.e. facility of molding the fresh concrete). This workability was measured by the slump test [9]. Compressive strength was then determined by ASTM C873 [10] for 3, 7, and 28 days.

The following conditions were tested for both temperatures of calcinations of the clay:

1. Without aggregate immersion
2. With aggregate immersion
3. With addition of additive

Concrete with gravel was used as a parameter because this is the most widely used aggregate in the region. This concrete was produced with and without the use of additive.

4.3 DISCUSSION AND RESULTS

4.3.1 Results of Physical and Mineralogical Characterization of Soil Samples

Table 1 indicates the results of the physical characterization of soil samples 1 and 2. This table shows the high plasticity of Soil 1, which prevented moisture from forming, and thus it was necessary to add more water to make the dough workable. This step indicates that the clay requires a moisture content of at least 55% to be formed. This amount of water directly affects the drying step and increases the risk of defects caused by shrinkage or crack formation and extension of drying time. It should be emphasized that the reasonable values of moisture for extruding a ceramic body using materials collected in Brazil are between 22 and 24% when using a vacuum extruder [11].

Table 1 Physical characterization of soil samples.

Samples	Clay (%)	Silt(%)	Sand (%)	Plastic Limit (%)	Liquid Limit (%)	Plasticity Index
Soil 1	43.96	50.27	5.77	55	86	31
Soil 2	43.53	44.22	12.27	28	50	22

The liquid limit (LL) is the moisture content at which a ceramic body begins to behave like liquid and is related to the adsorbed water associated with clay, corresponding to the maximum moisture that still allows it is molding [18]. Thus, the shaped ceramic body with moisture around the (LL) produced a very porous aggregate because the excessive amount of water was removed during drying and the pores previously occupied by water were empty.

Regarding composition of the soil, it can be observed in Table 1 that the content of clay and silt is similar in both soils, with different sand content. The lower sand content found in Soil 1 certainly contributed to the higher value of plasticity in this soil as compared to Soil 2. Soil 2 had a lower plastic limit but still within the parameter of 15%, which indicates the minimum moisture content for clay to still be conformed while minimizing the risk of cracks.

The lowest PI found in the soils used in this study was 22%; despite the limit previously established for the production of SACC, according to [11], a clay or clay mixture with a plasticity index (PI) greater than 15% can be used.

The results of X-ray diffraction, as seen in Figure 3 and 4, which agree strongly with results showed for calcined clay in previous studies [12, 13], indicated the presence of kaolinite and quartz. Neither of the soils was found to contain expansive clay minerals such as montmorillonite. However, there are characteristic diffraction peaks of illite, which belongs to the group of minerals found in clay and has essentially the same structure as mica, being less expensive than montmorillonite. In both soils the incidence of characteristic peaks of kaolinite was high, which is expected for common clay [12]. The low plasticity found in Soil 2 was due to the fact that there was a high percentage of non-plastic mineral such as, mica (muscovite) and goethite.

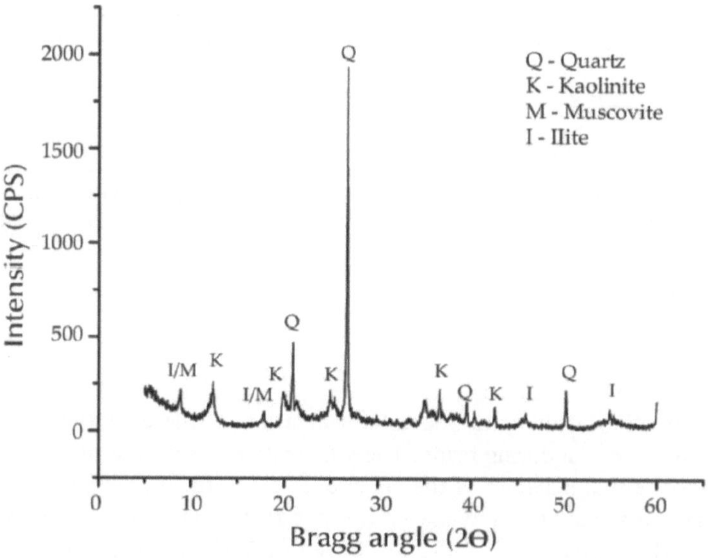

FIGURE 3 X-Ray Diffraction Analysis of Soil 1.

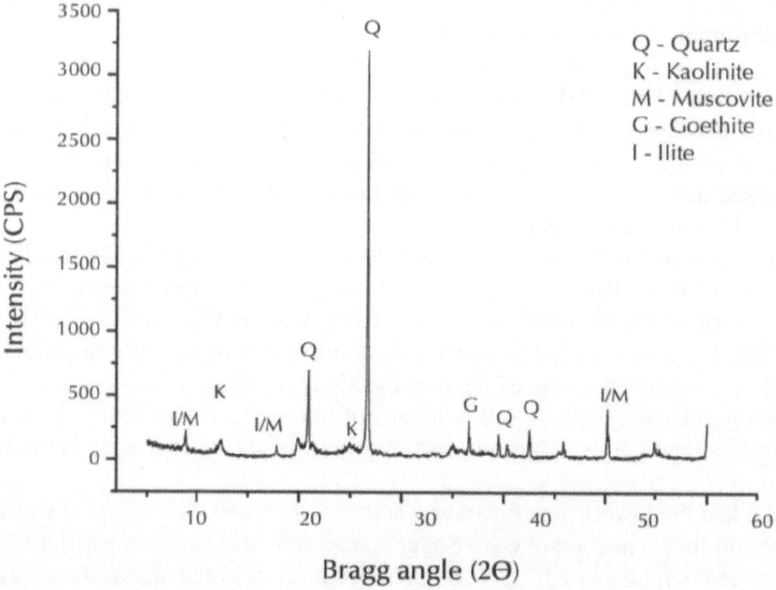

FIGURE 4 X-Ray diffraction analysis of soil 2.

4.3.1.1 Chemical Composition and Sintering of the Soil Samples

As show in Table 2 Soil 1 had a higher content of alumina, which indicates the presence of a higher percentage of clayey minerals. This percentage justifies the high level of consistency found for this soil. These two soils were seen to contain a value exceeding 5% of iron oxide, responsible for the reddish color of the ceramic material before and after burning. The concentrations of silica, aluminum and iron oxide are characteristic of the raw material used in the production of red tile, which means that the same facilities used in red tile manufacture can be used to produce SACC.

TABLE 2 Chemical composition of soil samples.

Samples	SiO_2	Fe_2O_3	Al_2O_3	MgO	CaO	TiO_2	Na_2O	K_2O	MnO	LOI*
Soil 1	66.0	6.2	16.11	1.5	1.1	0.9	1.4	2.0	0.1	4.7
Soil 2	59.56	7.3	14.55	1.21	0.92	1.2	1.5	2.19	0.11	11.42

* L.O. I. – Loss on Ignition.

The vitrification of ceramic bodies is a measure of the evolution of the microstructure of the material during firing. The vitrification curve used in ceramic material illustrates the thermal behavior of the material as a function of temperature [14]. This curve shows the variations of WA and linear retraction (LR) as a function of firing temperature. This type of curve indicates the temperature where the material begins to ceramicize, corresponding to the point of intersection between the curve of absorption and shrinkage. The intersection point is the starting point for the vitrification of the material, characterized by chemical reactions between the interfaces of grains where two grains gradually unite to form one.

As can be seen in Figures 5 and 6, soils 1 and 2 began to be vitrified starting at 1,125°C. From 850 to 950°C l variation is observed in the percentage of WA and a degree of dimensional stability of parts, due to low linear shrinkage. From 1,000°C an increase in shrinkage occurs, going from slightly over 1% to around 7%, indicating an increased liquid phase, derived from the reaction of the flux elements, reducing the WA from 15% to less than 6%.

Soil 1 shows a flux content (Fe_2O_3 + CaO + MgO + Na_2O + K_2O) of around 12.20%, while in Soil 2 this value is about 13.12%, which in theory would mean that the latter should be denser. On the other hand, considering that densification is directly related to porosity, a high loss on ignition of 11.4% contributed significantly to an increase in the porosity and thereby reduced the densification.

The high plasticity in Soil 1 is evidence of fine particle size, which increases the degree of compaction (packing) in fresh clay, thus resulting in a more dense ceramic material.

Soil 2 had the lowest specific gravity and a lower moisture level, so it was used as raw material for production of coarse aggregate calcined clay at 850 and 1,125°C. The choice of temperature of 850°C was made because, as noted above, furnaces in the region reach temperatures in the range from 850 to 950°C. However, at a temperature of around 1,125°C the ceramic mass for both soils reached the lowest value for absorption and the maximum value for linear shrinkage. This study examined the effect of these two temperatures in the aggregate structure of calcined clay. It should be noted that these aggregates were calcined in an oven for a period of 1 hr after reaching the top temperature.

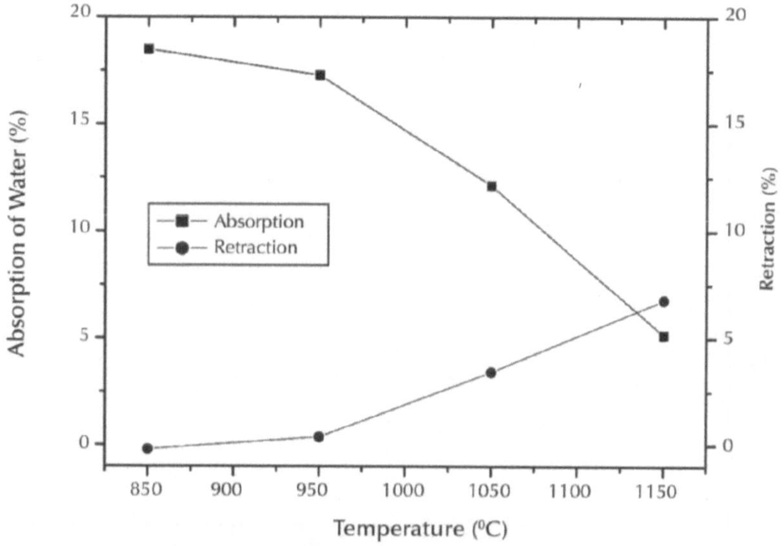

FIGURE 5 Vitrification curve of Soil 1.

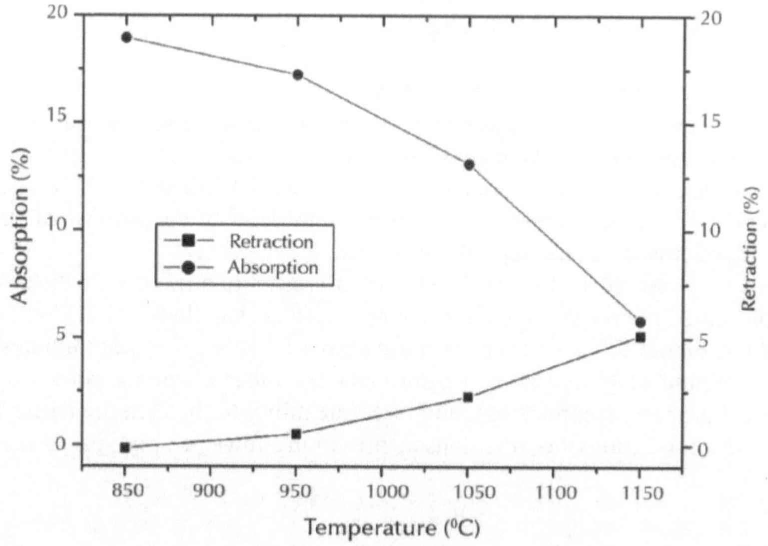

FIGURE 6 Vitrification curve of Soil 2.

Figure 7 presents the results of the specific gravity analysis of fired soil specimens. It is observed that up to 950°C, the densification of soils 1 and 2 is less than 2 g/cm³ and at 1,125°C the density reaches values of 2.78 and 2.04 g/cm³, respectively.

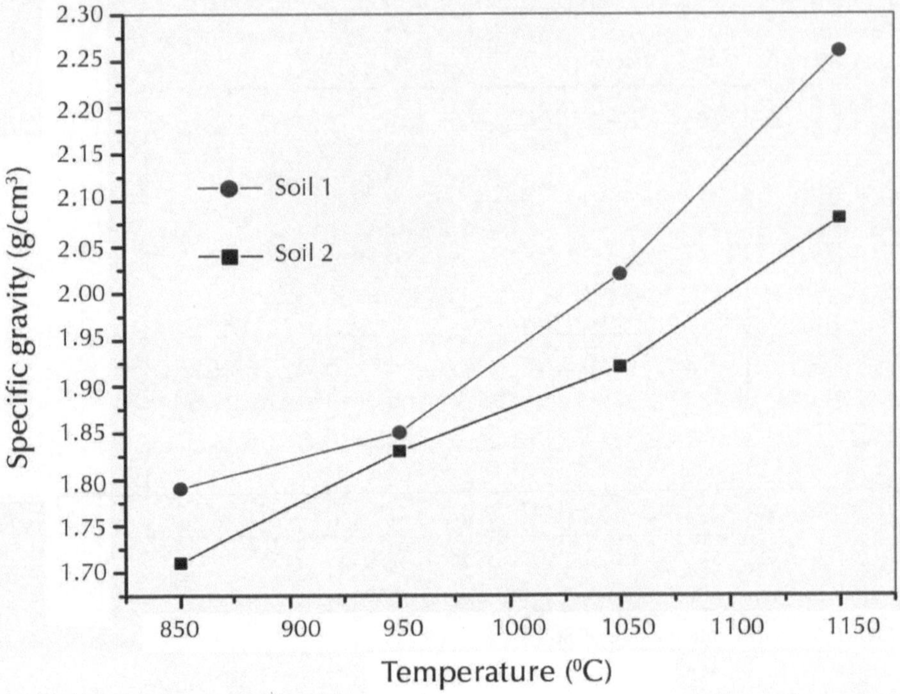

FIGURE 7 Specific gravity vs. temperature.

4.3.1.2 Morphological analysis of SACC

With calcination, the clay acquires stability and mechanical strength as a result of some physical and chemical changes.

Small samples of SACC, produced at temperatures of 850 and 1,125°C, were collected for the scanning electronic microscopy analysis, with the purpose of verifying the internal microscopic structure of this aggregate.

It can be observed in Figure 8 that the microscopic structure of the SACC at 850°C presented microcracks and more empty spaces than that of SACC at 1.125°C as shown in Figure 9. This occurs because above 1.000°C, silico-aluminates create a certain amount of glass, which agglutinates the other elements, providing more resistance, hardness, compactness, and impermeability to the ceramic parts. Thus at 1,125°C, the clay grains became denser, presenting fewer empty spaces inside the clay.

FIGURE 8 Microscopy of SACC made at 850°C (Beam voltage = 20 kV; Magnification = 27 times, and Scale bar = 500 mm).

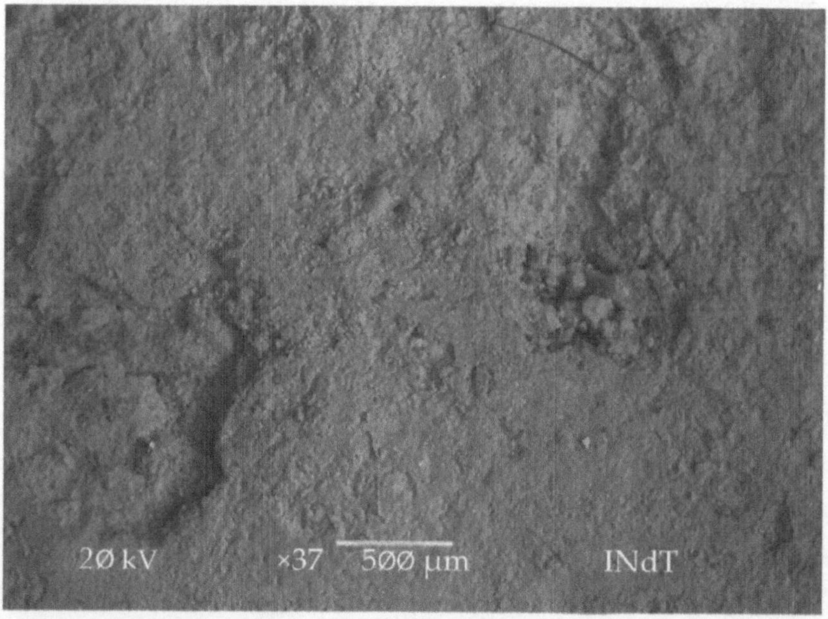

FIGURE 9 Microscopy of SACC made at 1,125°C (Beam voltage = 20 kV; Magnification = 37 times, and Scale bar = 500 mm).

4.3.1.3 Coarse Aggregate Characterization Results

For the application of SACC in concrete, physical parameters were obtained through tests using ASTM C1138, ASTM C127, ME 225/94 [15-17]. The results are presented in Table 3.

TABLE 3 Physical characterization of coarse aggregate.

Characterization	Aggregate		
	SACC at 850°C	SACC at 1125°C	Gravel
Abrasion	52%	48%	22%
Loss weight after boiling	0.1%	less than 0.1%	--
Specific mass	1.70 g/cm³	1.94 g/cm³	2.63 g/cm³
Unit mass	1.08 kg/dm³	1.16 kg/dm³	1.86 kg/dm³
Absorption	18.94%	12.5%	1.22%

Gravel, used as natural aggregate, has a maximum size of 19 mm, with more than 37% of the material being inferior, at less than 4.8 mm (Figure 10). The SACC, on the other hand, was made with a maximum size of 12.5 mm, so it could present a behavior similar to conventional aggregate. It is lower limit of 4.8 mm was established due to the limitations of production molds. It should be noted that gravel has a high percentage of fine particles, while SACC had the lowest voids when mixed with a percentage of 55, 20, and 25% of aggregates of the dimensions 4.8, 9.5, and 12.5 mm respectively.

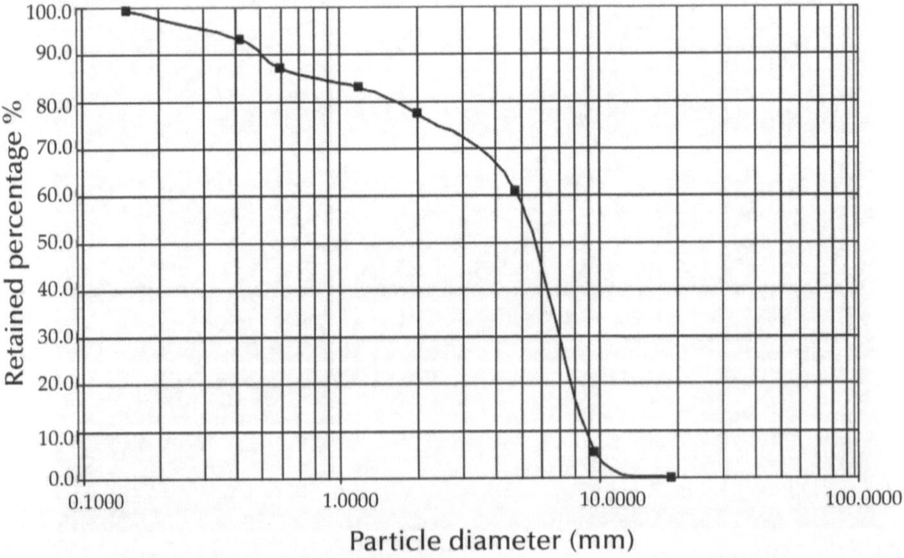

FIGURE 10 Granulometry curve of gravel.

The abrasion of SACC produced at 850°C was above the limit established by ASTM C 1,138 [15], which are 50%. For the SACC produced at 1125°C, however, the value presented for the abrasion test was in accordance with this norm. The abrasive fret is important when the concrete is intended for paving but it is not a determining factor for applications in structural concrete.

The mass loss after boiling reached levels below 1%, while ME 225/94 [17] recommends that levels should not exceed 6%.

Regarding the absorption of water, the SACC at 850°C had values above 18%, while for the aggregates burned at 1125°C the value was 12.5%. This reduction of 5.5% in the WA will have to come at a cost in terms of energy input, since the kilns in the region's ceramics industry work at temperatures of only 850–950°C.

Even the temperature was raised to 1,125°C, the SACC still had very high porosity compared to the gravel, which had a porosity of 1.22%. It is possible that the high WA was due to the use of the manual mill, which required the ceramic body to have a moisture content of around 28%.

Considering that the molding water was eliminated during firing, leaving the pores empty, the aggregates at the end of the process showed high porosity.

4.3.1.4 Concrete without Immersion of Aggregate in Water

All mixes of SACC produced at 850°C presented a high water/cement ratio, as can be observed in Table 4. This proved that due to the high porosity of SACC, the mixes absorbed water from the concrete mixture. In addition, SACC also reduced the specific gravity of the concrete when compared to concrete made with gravel. The concrete strength for different cement consumption at 28 days reached just over 22 MPa, above the limit of 20 MPa established by ASTM C873 [10] for structural concrete. It is noteworthy that the concrete obtained with a consumption of 335 kg/m³ had the best strength / cement consumption ratio.

Table 4 Concrete without immersion in water.

Aggregate	**Mix**	**Cement consumption (Kg/m3)**	**Slump (cm)**	**Concrete specific gravity (Kg/m3)**	**Strenght (MPa)** 28 days	**Water/Cement ratio**
SACC at 850° C	1	335	7.8	2114	22.79	0.85
	2	352	8.2	2108	22.41	0.81
	3	385	9.0	2103	22.55	0.75
SACC at 1125°C	1	336	7.5	2113	26.44	0.70
	2	361	10.0	2130	26.62	0.73
	3	386	9.2	2132	27.86	0.68
Gravel		351	8.0	2324	28.12	0.55

For concrete produced with SACC made at 1125°C, the data in Table 3 show that the water/cement ratio of the concrete was less than the aggregate calcined at 850°C. This reduction could be explained by a decrease in porosity from 18 to 12.5% caused by internal changes of the aggregate during firing. Thus, there was less absorption of the water from the concrete mixture, which improved the workability and concrete strength. The strength of the concrete at 28 days was higher as compared with the SACC concrete produced at 850°C. It is also possible to observe a small variation in the specific gravity of the concrete because it reached values above 2,130 kg/m³.

For SACC concrete produced at both temperatures the water/cement ratios were high (on average about 0.80 and 0.70) when compared to that for gravel. This proves the high porosity of SACC, which should soak up some water from the concrete mixture.

Both concretes using SACC made at 850 and 1125°C presented high compressive strength and high water/cement ratio, which were needed to offset the decline in workability. These results are in accordance with those presented by Khalaf [5] in which concrete was produced using crushed brick as the coarse aggregate without the material being immersed in water before mixing.

4.3.1.5 Concrete with Immersion of Aggregate in Water

The SACC was immersed in water for 24 hr before mixing (Table 5).The effect of this pre-wetting with SACC made at 850°C was notably on the water/cement ratio, which was reduced to values below 0.5, compared to concrete with SACC made at 850°C without pre-wetting, which reached water/cement ratio averages of 0.8. This pre-wet concrete achieved mean workability of 7 cm of slump. The specific gravity of the concrete did not change significantly with immersion in water. The immersion of the aggregate also improved the workability without affecting the resistance because there was not need for the SACC to soak up some of the water from the concrete mixture. The presence of water promotes the hydration of the transition zone, making the concrete stronger. Thus, it was possible to obtain a compressive strength exceeding 28 MPa at 28 days, much higher than that obtained with the aggregate not immersed and in close proximity to the value for the concrete made with gravel.

Table 5 Concrete with immersion in water.

Aggregate	Mix	Cement consumption (Kg/m³)	Slump (cm)	Conrete specific gravity (Kg/m³)	Strenght (MPa) 28 days	Water/Cement ratio
	1	345	7.0	2096	27.24	0.47
SACC at 850°C	2	368	6.0	2110	27.95	0.45
	3	388	8.0	2120	28.85	0.43
	1	334	9.2	2119	28.93	0.54
SACC at 1,125°C	2	355	8.5	2096	30.24	0.51
	3	379	7.5	2110	31.15	0.48
Gravel		351	8.0	2324	28.12	0.55

For SACC made at 1,125°C, the immersion in water generated a concrete with a lower water/cement ratio. The highest value of the water/cement radio obtained was

0.54, with a slump of 9.2 cm. The compressive strength was above 30 MPa for most of the mixes.

In both concretes produced with SACC made at 850 and 1125°C, improvement of workability, compressive strength, and water/cement radio corroborates with results presented by Khalaf [5], in a study in which crushed brick aggregate was immersed in water for 30 min before being used in concrete.

4.3.1.6 Concrete with Additive

In order to improve workability a calcined clay aggregate concrete mix was designed and produced with the addition of a superplasticizer admixture. The additive was used in the proportion of 310 ml for each 1 kg of cement.

The aggregate was not pre-wetted before mixing because it was expected that the superplasticizer would improve the workability and allow the pre-wetting procedure to be avoided. However, the water/cement ratio was only slightly smaller when compared with SACC no submitted to prior immersion. It is also evident in Table 6 that the water/cement ratios of concrete with SACC at 850 and 1125°C could be reduced, minimizing the effects of porosity.

TABLE 6 Concrete with additive.

Aggregate	Mix	Cement consumption (Kg/m³)	Slump (cm)	Conrete specific gravity (Kg/m³)	Strenght (MPa)	Water/Cement
					28 days	ratio
	1	334	6.9	2099	26.58	0.77
SACC at 850° C	2	352	7.5	2117	27.01	0.75
	3	370	7.2	2017	27.19	0.68
	1	332	6.8	2112	31.12	0.69
SACC at 1125°C	2	345	7.0	2115	32.65	0.67
	3	382	9.0	2090	33.07	0.61
Gravel		354	7.3	2302	34.20	0.46

It was possible to have a slump of around 8 ± 1 cm with the use of the additive in both concretes with SACC and with gravel.

The concrete resistance with gravel was superior to that with SACC, as was expected since this material is less porous and does not absorb significant amounts of the water from the concrete mixture. For this reason, the additive was more effective with the gravel, reducing the water/cement ratio from 0.55 to 0.46 and thus improving the final resistance of the hardened concrete.

It must be emphasized that the concrete produced with the SACC at 1,125°C has a specific gravity lower than that produced with gravel. In other words, this product could replace gravel in all of its applications and will have the advantage of generating lighter building materials than those produced with gravel.

4.4 CONCLUSION

Soil 2 used in this study, from Manacapuru, presented lower specific gravity and moisture molding when used as raw material for the production of SACC coarse aggregate at 850 and 1125°C.

This soil also presented low plasticity, providing a ceramic body with a lower percentage of water in its conformation. This directly affected the drying time and reduced defects caused by retraction.

On the behavior under calcination, the soils presented high absorption values until 1,000°C, and some dimensional stability with small volumetric contractions. In the production of ceramics in Iranduba and Manacapuru, more attention needs to be paid to porosity.

The calcined aggregates at 850°C presented WA of 18%, while at 1,125°C this percentage fell to 12.5%. However, increasing the amount of energy used in order to raise temperature from 850 to 1,125°C in a ceramic industrial kiln will have a significant environmental impact. In addition, this change of temperature in the SACC production process does not produce a significant decrease in the WA rate.

The electron microscopy scanning analysis made on SACCs at 850 and 1,125°C confirmed the increase of porosity with a decrease of calcining temperature.

The specific gravity of SACC aggregate increased with the increase in the calcinations temperature because the decrease of porosity increased densification.

The SACC's specific gravity was lower than that for gravel, which contributed to the reduction of the specific gravity of the concrete.

The concrete made with SACC not immersed in water before casting reached a compressive strength above the limit of 20 MPa established by ASTM C873 [10] for structural concrete at 28 days. These results were achieved for cement consumption of 335 kg/m^3 (SACC at 850°C) and 336 kg/m^3 (SACC at 1,125°C). In this condition, both SACC concretes showed concrete specific gravity less than that obtained for gravel. Nevertheless, the water/cement ratio was high in both of the SACC concretes due to the high porosity of SACC, which absorbed water from the concrete mixture, creating the need to increase the amount of water in the mix to compensate for the loss of workability.

The immersion of the aggregate before mixing improved the most important properties of SACC concretes. This procedure also improved the workability without affecting the concrete specific gravity, which did not change significantly, or the concrete compressive resistance because there is no need for the SACC to soak up some of the water from the concrete mixture, which promotes the hydration of the transition zone, making the concrete stronger. Considering only the lower consumption cement, 345 kg/m^3 (SACC at 850°C) and 334 kg/m^3 (SACC at 1,125°C), the compressive strength not only reached values higher than the limit of 20 MPa established by ASTM C873 [10] for structural concrete at 28 days but also similar (27.24 MPa for SACC at 850°C) to those obtained for the gravel (28.12 MPa) as above (28.93 MPa for SACC at 1125°C). The specific gravity of the concrete did not change significantly with immersion in water.

The immersion of SACC can be avoided through the production of concrete with the addition of a superplasticizer admixture. Therefore, this procedure, as expected, improved the workability and decreased the water/cement ratios of the SACC concrete. Furthermore, it maintained practically the same level of compressive strength when compared with SACC concretes submitted to the process of immersion in water.

All of the SACC concretes reached a compressive strength above the limit of 20 MPa established by ASTM C873 [10], even considering only the lower consumption cement. The strength, workability, and water/cement ratio could be improved by using the process of prior immersion of SACC or adding a superplasticizer admixture. The SACC made at 1,125°C presented a technical advantage as compared to the SACC made at 850°C; however, the latter presented economic advantage. Therefore further studies are needed to determine the relative advantages of incorporating the process of immersion, adding superplasticizer or producing SACC at a higher temperature.

4.5 ACKNOWLEDGMENTS

This research was funded by Fundo Setorial de Infra-Estrutura (CT-INFRA) through MCT/CNPq and Fundação de Amparo a Pesquisa do Estado do Amazonas (FAPEAM). Raimundo Kennedy Vieira thanks to the CAPES for the program PRO-ENGENHARIA.

KEYWORDS

- **Calcined Clay**
- **Liquid Limit**
- **Mineralogical Analyses**
- **Montmorillonite**
- **Synthetic Aggregate Calcined Clay**
- **Vitrification Curv**

REFERENCES

1. Santos, P. S. *Tecnologia de Argilas. 1st edn., Vol. 2,* Editora Edgarg Blücher, São Paulo, Brazil (1975).
2. Fouad, M. K. and DeVenny, A. S. Properties of New and Recycled Clay Brick Aggregates. *J. Mater. Civ. Eng.*, **17**, 457464 (2005).
3. Toledo Filho, R. D. and Gonçalves, J. P., Americano, B. B., and Fairbairn, E. M. R. Potential for use of crushed waste calcined-clay brick as a supplementary cementitious material in Brazil. *Cem. Concr. Res.*, **37**, 13571365 (2007).
4. Frias, M., Rodriguez, O., Vegas, I., and Vigil, R. Properties of Calcined Clay Waste and its Influence on Blended Cement Behavior. *J. Am. Ceram. Soc.*, **91**, 12261230 (2008).
5. Khalaf, F. M. Using Crushed Clay Brick as Coarse Aggregate in Concrete. *J. Mater. Civ. Eng.*, **18**, 518526 (2006).
6. Neville, A. M. *Concrete Properties. 1st edn., Publisher PINI,* São Paulo, Brazil (1997).

7. CPRM. Serviço Geológico do Brasil—Geologia e recursos minerais do Estado do Amazonas. (Geology and mineral resources of the State of Amazonas). Manaus, Brazil, CPRM, CD-ROM. Portuguese (2006).
8. ASTM Standard ASTM C494/C494M, 2010, *Specification for Chemical Admixtures for Concrete*, ASTM International, West Conshohocken, PA (2003), doi: 10.1520/C0494_C0494M-10, www.astm.org.
9. ASTM Standard ASTM C143/C143M, 2010, *Slump of Hydraulic-Cement Concrete*, ASTM International, West Conshohocken, PA (2010)_ doi: 10.1520/C0143_C0143M-10, www.astm.org.
10. ASTM Standard ASTM C873/C873M, *Compressive Strength of Concrete Cylinders Cast in Place in Cylindrical Molds*, ASTM International, West Conshohocken, PA (2010) doi: 10.1520/C0873_C0873M-10, www.astm.org.
11. Vieira, C. M. F., Souza, E. T. A., and Monteiro, S. N. Ceramic bodies for roofing tiles: Characteristics and firing behavior. *Cerâmica*, **49**, 245250 (in Portuguese) (2003).
12. Baronio, G. and Binda, L. Study of the pozzolanicity of some bricks and clays. *Constr. Build. Mat.*, **11**(1), 4146 (1997).
13. Said-Mansour, M., Kadrib, E., Kenaia, S., Ghricic, M., and Bennaceur, R. Influence of calcined kaolin on mortar properties. *Constr Build Mater.*, **25**(5), 22752282 (2010).
14. Sánchez-Muñoz, L. C. S. S., Paskocimas, C. A, Cerisuelo, E. L. E., and Carda, J. B. Selection of raw materials in the development of ceramic bodies compositions. *Cerâmica*, **48**, 108113 (in Portuguese) (2002).
15. ASTM Standard ASTM C1138, 1997 *Abrasion Resistance of Concrete (Underwater Method)*, ASTM International, West Conshohocken, PA (1997) doi: 10.1520/C1138-97, www.astm.org.
16. ASTM Standard ASTM C127, 2007, *Density, Relative Density (Specific Gravity), and Absorption of Coarse Aggregate*, ASTM International, West Conshohocken, PA (2007) doi: 10.1520/C0127-07, www.astm.org.
17. DNER, ME 225/94 *Aggregate synthetic clay calcined: Determination of mass loss after boiling*. National Department of Transport Infrastructure, Rio de Janeiro, RJ, (in Portuguese) (1994).
18. ASTM Standard ASTM D4318, 2010, *Liquid Limit, Plastic Limit, and Plasticity Index of Soils*, ASTM International, West Conshohocken, PA (2010) doi: 10.1520/D4318-10, www.astm.org.

5 Recent Advances in Rubber Recycling a Review

S. K. Kumar, S. Thomas, and Y. Grohens

CONTENTS

5.1 INTRODUCTION

Waste disposal management is one of the serious problems which mankind is facing in this century. Management of solid waste is especially important with most materials. Polymeric materials (plastics and rubbers) comprise a steadily increasing proportion of the municipal and industrial waste going into landfill in many countries, even though global plastics production is estimated to have fallen from 245 mt in 2008 to around 230 mt in 2009 as a result of the economic crisis. Development of technologies for reducing polymeric non-recycled waste, which are acceptable from the environmental standpoint, and which are cost-effective, has proven to be a difficult challenge due to complexities inherent in the reuse of polymers. Europa for instance has introduced several policies and regulations for plastic recovery and recycling targeted at key industrial sectors (Waste Framework Directive, 2008/98/EC). The plastics recycling rate was 21.3% in 2008, helping to drive total recovery (energy recovery and recycling) it should be enhanced to 51.3% whereas the US recycling rate was only 8% in 2010. Establishing optimal processes for the reuse/recycling of various elastomeric materials thus remains a worldwide challenge.

Rubber, produced from natural or synthetic origin is a technologically important material. Natural Rubber is obtained from the tree *Havea Braziliensis* and synthetic rubbers by polymerization reaction. The synthetic rubbers include neoprene, Buna rubbers, and butyl rubber. The industrial application of rubbers is unlimited due to its unique physical and chemical properties. The most important primary (intermediate) rubber products include concentrated latex (raw material for dipped products such as medical gloves and condoms), block rubber (raw material for high viscosity products such as soles and belts), and ribbed smoked sheet rubber (raw material for vehicle tires and industrial rubber parts). The applications of rubber include medical, industrial, automobile, space science, and in short all phases of life. Because of their excellent elongation and recovery properties, rubber is used in textile industry also. The contribution of rubber industry to the world economy is represented by Figure 1 that shows a tremendous increase in use of rubber products which necessitates relevant recycling processes and policies.

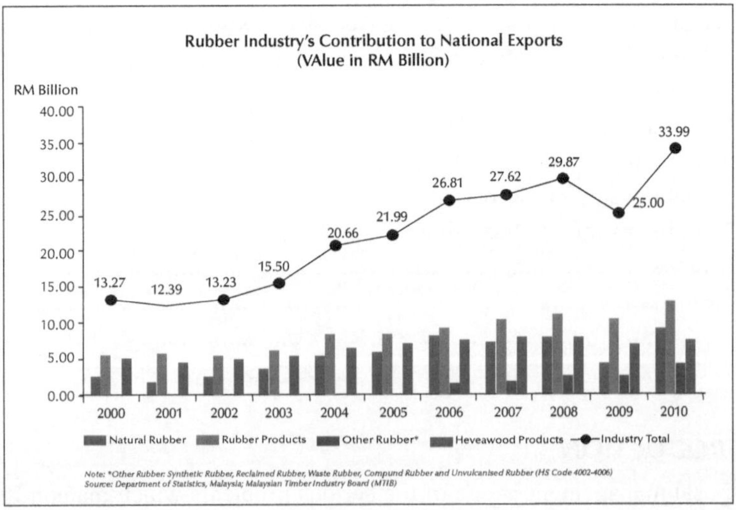

FIGURE 1 Rubber Industry's contribution to National Exports http://www.lgm.gov.my/.

Even though rubber products exist in wide varieties, tire industry stands first out of all these applications. Because of this reason, rubber recycling is often referred to tire recycling. Tires are among the largest and most problematic sources of waste, due to the large volume produced and their durability. The high energy absorbing capacity of rubber affects the tire disposal/recycling problem in two ways. First, it is extremely difficult to disassemble tires into small, easy to handle pieces. The shape of the entire tires does not allow efficient waste transportation. Another problem is due to the vulcanization effect. The vulcanization is done on rubber during tire manufacture in order to improve their mechanical and thermo elastic properties. This causes in the formation of strong sulfur bridges between the hydrocarbon chains and thus making it impossible to melt and reshape as can be done with thermoplastic materials. The most common method for tire disposal is incineration or landfill. But both methods are found to be harmful to the environment. Also the disposal of scrap tires in landfills has proven to be a problem because whole tires can "float" to the surface and break the cap of the landfill. It is claimed that a buried tire with ground material on top of it trickle down around its sides, until there is more pressure forcing the tire up than is keeping it down. Even if this is a slow process, there is a chance for the tires could come up to the surface, after years. Small quantities of rubber might be beneficial to plants by increasing the porosity, but as the quantity increases, sensitive plants will likely be adversely affected [1].

Though, the waste tires create a lot of problems, they are one of the most reused waste materials, as the rubber is very resilient and can be reused in other products. The recycling of rubber or tire can be defined as the process of recycling vehicles tires that are no longer suitable for use on vehicles due to wear or irreparable damage. Recycled rubber can be generalized as any rubber waste that has been converted to an economically useful form, such as reclaimed rubber, ground rubber or reprocessed synthetic rubber [2]. Tire industry usually considers four R's which are Reduction, Reuse, Recycling, and Recovery that is in general, a priority order for decreasing the amount of waste should be: (1) reduction of consumption; (2) reuse of the product; (3) recycling of materials; (4) energy recovery; and (5) as a last possibility, deposition of the waste. Figure 2(a) gives an account of percentage of recycling rates of selected products in the year 2010. Among various products, the recycling rate of tire is found to be only 35.5%, which has to be improved much for a better future. In the recycled tire components, 41% comprise of natural as well as synthetic rubber, as per the Figure 2(b). This depicts the importance of an effective study in the area of rubber or tire recycling. The significance of recycling of waste rubber in protecting the environment and conserving energy is discussed in detail in this chapter. Also, various kinds of recycling approaches to waste rubber are summed up, such as reclaiming energy as fuel, reuse of the products of thermal decomposition, cleaning of leaking oil, reuse after simple modification, regenerative rubber and powdered rubber. The crumb rubber, produced from scrap tires can have wide range of particle sizes and quality levels. The sale of variously sized crumb products to different end-user markets and potential sales of scrap metal and fiber contained within the tires can add to the country's economy. So, the general demand has been increasing, and submarkets for crumb products are growing in size and variety. However, the optimistic expectations of potential investors and government agencies contrast sharply with the experiences of many current and former producers. In addition, numerous projects have been conducted on replacement of aggregates by crumb rubbers but scarce data are found on cementitious filler addition in the literature. This chapter also examines the engineering economics of crumb rubber facilities and their effective applications.

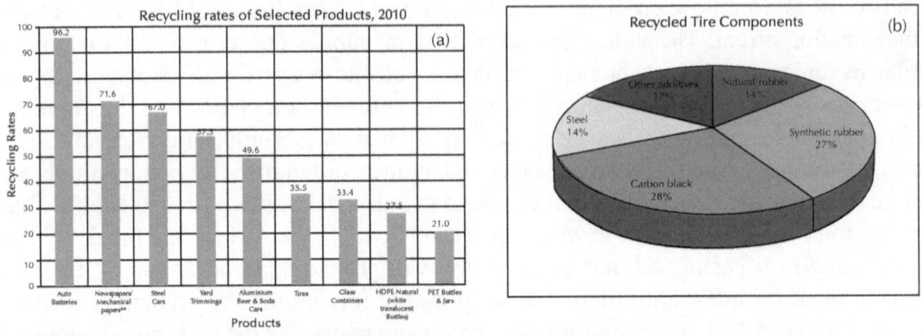

FIGURE 2 (a) Recycling rates of different products (b) recycled tire components.
http://www.epa.gov/osw/nonhaz/municipal/
http://www.accuval.net/insights/industryinsights/detail.php?ID=114

The recycled rubbers can be effectively utilized in several ways. Reports show that recycled low density polyethylene (R-LDPE) has been reactively compatibilized with butadiene rubber (BR) by using small additions of reactive polyethylene copolymers and reactive BRs to produce thermoplastic elastomers (TPEs) [3]. In a similar way, Ground tire rubber (GTR) was effectively embedded into polyethylene (PE) and polyethylene/ethylene-vinyl acetate copolymer (PE/EVA) matrices [4]. The gamma irradiated tire inner tube wastes made of butyl rubber and commercial butyl rubber crumbs devulcanized by conventional methods are used to prepare butyl based rubber compounds [5]. Also it is found that the waste tyre rubber (WTR) can be reused for oil absorptive material production [6]. Due to the increasingly serious environmental problems caused by waste tires, the feasibility of using elastic and flexible tire–rubber particles as aggregate in concrete is found to be an effective way of waste disposal. Tire–rubber particles can be made to tire chips and crumb rubber, and a combination of these two, was used to replace mineral aggregates in concrete. Ultrasonic analysis reveals large reductions in the ultrasonic modulus and high sound absorption for tire–rubber concrete. Rubberized concrete is recommended for highway barriers. It was observed that rubberized concrete's acquired competency to absorb energy resulted in decrease in damages and injuries during the collision of vehicles with barriers. Also the impact resistance of rubberized concrete was higher, and it was particularly evident in concrete samples aggregated with thick rubber. Researches on rubber waste management are going on in all parts of the world. Bruno et al. described the development and implementation of a regulation based on the extended producer responsibility (EPR) concept towards tyre waste in Brazil [7], the birthplace of natural rubber. They suggested that, although Brazil tried to catch up with industrialised countries by importing a foreign model, the outcomes were not satisfactory. Failure is associated with a partial implementation of the EPR concept and with the limited institutional capacity of the federal environmental agency [8]. Jawjit et al. tried to quantify greenhouse gas emissions associated with the production of fresh latex and primary rubber products in Thailand and formulated certain methods to reduce these emissions [9]. The recycling is also expressed by another term Reclaiming, which is defined as the conversion of a three dimensionally interlinked, insoluble and infusible strong thermoset polymer toa two dimensional, soft, plastic, more tacky, low modulus, processable, and vulcanizable essentially thermoplastic product simulating many of the properties of virgin rubber [10].

Despite the fact that more and more methods and solutions are used in the recycling of rubbers, there are still some problems, especially in the recycling of cross-linked rubbers. Usually the biggest problem is the lack of compatibility between the cross-linked rubber and the matrix, which is used for blending with it. This chapter discusses the challenges in rubber recycling and the recent advances in the recycling technology

5.2 METHODS OF RECYCLING

A tire is a composite of complex elastomer formulations, fibers and steel/fiber cord. Tires are made of plies of reinforcing cords extending transversely from bead to bead, on top of which a belt is located below the thread. So the recycling of tires can produce three components mainly, elastomer, carbon black, and steel wire. These elastomers are a source of crude oil as represented by Figure 3.The important ways to recycle the tires include (1) use of tire rubber in asphaltic concrete mixtures; (2) incineration of tires for the production of steam and (3) reuse of GTR in a number of plastic and rubber products [11].

FIGURE 3 A model for tire recycling process.

General methods of recycling of rubbers/tires can be broadly classified in to physical methods, chemical methods, and biological methods based on the way of processing. All the methods can be summarized as shown in Figure 4.

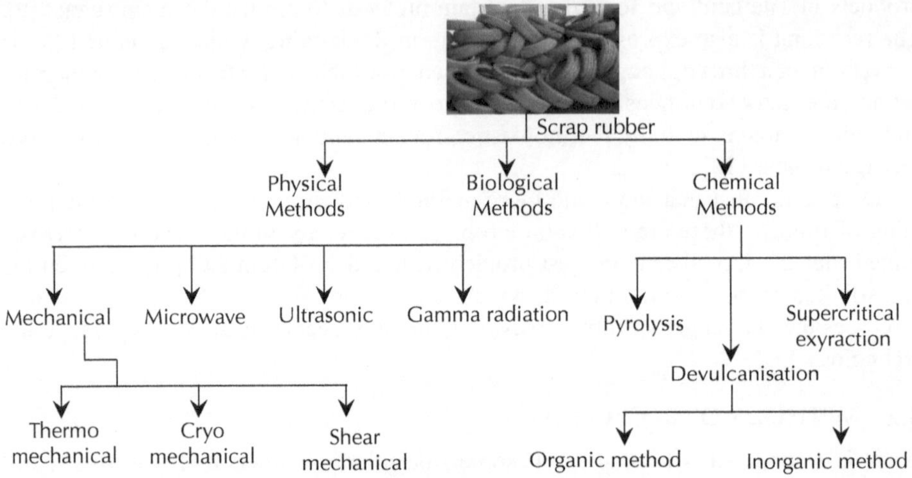

FIGURE 4 Different recycling methods of rubber.

Since the three-dimensional vulcanized rubber is unable to flow during the manufacturing processes, it should be devulcanised first for tire recycling. The process of devulcanization involves the cleavage of cross-linking sulfur bonds in rubber vulcanizates, without breaking the polymer chain bonds [12, 13]. Devulcanization is recognized as the best way of utilizing rubber waste since it assumes renewal of the original chemical formula of elastomers and provides a possibility of recovering elastomers from rubber vulcanizate waste. During this process the sulfur cross-links breaks and the elastomer chains become free and this can be processed easily.

5.2.1 Physical Method

This method includes treating the rubber molecules without the use of chemical reagents.

5.2.1.1 Mechanical Treatment

Mechanical recycling method involves milling of crumb rubber in two-roll mill. The molecular weight of the rubber decreases drastically due to mechanical shearing. High-shear mechanical milling helps to pulverize the polymer materials effectively with improved surface activation. This method has a variety of advantages over other de-cross-linking methods. The process does not require use of any chemical agent for de-cross-linking and does not generate byproducts. It minimizes the use of organic solvents and also produces reclaimed rubbers with superior mechanical properties. This is an energy efficient process since it is operated at ambient temperature. De et al. [14] reported the mechanical reclaiming process of vulcanized NR. They studied the curing characteristics and mechanical properties of composites of reclaim rubber (RR) and pure rubber. The very high Mooney viscosity of the reclaimed rubber indicates the low plasticity of rubber due to the presence of higher percentage of cross-links. They observed an increase in cure rate with increase in the content of reclaimed rubber but the scorch time, optimum cure time, and reversion resistance are decreased. The mechanical properties like the modulus, abrasion loss, compression set, and hardness increases with the increase in proportion of reclaim rubber in the blends, whereas the tensile strength, elongation at break, tear strength, resilience and flex resistance decreases. Thus the cross-link density of NR/RR (25/75) blend is observed to be high with low tensile strength and flex properties. However, they could not explain the extent of reclaiming, molecular weight of the sol fraction, and the influence of milling parameters on the Mooney viscosity.

The mechanical devulcanization processes include thermo mechanical, cryo mechanical, and shear mechanical reclamation. Thermo mechanical reclaiming process involves the thermo mechanical degradation of the rubber vulcanizate network. In this process, first the vulcanizate is swollen in a suitable solvent and then transferred to a mill to make a fine powder and this powder rubber is revulcanized with curing ingredients. The products thus obtained show slightly inferior properties to those of the original vulcanizates. The significant amount of heat generated in the rubber, during this process is the disadvantage of this method. This excess heat can degrade the rubber and it has the potential danger of combustion if not cooled properly. Whereas the cryo-mechanical reclaiming process involves adding small pieces of vulcanized rubber into liquid nitrogen and then transferred to a ball mill and ground in presence of liquid nitrogen to form a fine powder. The particle size is controlled by the immersion time in the liquid nitrogen and by the mesh size of screens used in the grinding chamber of the mill. For less particle size, the cost of production is high. This method avoids the heat generation and there by the possible degradation that can happen in the rubber molecules. Norman et al. reported about the cryogenic studies for the recycling of tires as a promising future processing [15]. It has also been reported that using 5–10 phr cryogenically ground rubber in tire compounds have some economic advantage [16]. Maxwell [17] reported the recycling of GTR using a stator and rotor arrangement, in which the rubber is frictionally propelled by shear. They mixed previously reclaimed rubber and vulcanized rubber and the method is found to be useful as a substitute for conventional refining operation. Zhang et al. [18] developed an eco-friendly approach for the devulcanization of fluoroelastomer (FKM) scraps through mechanical shear milling. The recycling of FKM is important because of its high cost. The FKMs are used as high performance seal materials (engine oil seals, hoses, cables, etc.) in many industrial and space applications due to their excellent thermal, oil, and chemical resistance arising due to the C–F bonds. This strong chemical bond makes its recycling difficult by the conventional methods. So high shear mechanical milling is essential for the devulcanization of these materials. They observed a decrease in gel fraction from 97.8 to 79.7%, after 32 cycles of milling, indicating the occurrence of stress induced mechanical de-cross-linking of FKM. The reclaimed FKM exhibited excellent mechanical and thermal properties, indicating a strong potential for future applications. The tensile strength of FKM revulcanizates (6.6 MPa) is near to that of virgin FKM vulcanizates (7.9 MPa), and the elongation at break was increased from

337.1 to 368.7%.They also confirmed the structural change of FKM sol part before and after mechanical milling by Fourier transform infrared (FTIR) analysis and gel permeation chromatography (GPC) measurements.

5.2.1.2 Ultrasonication

Since the ultrasonic energy is very powerful, it can be utilized for the devulcanization process. These waves can create high frequency extension contraction stresses in rubber matrix and thus can break the C–S bonds. Pelofsky [19] in 1973 reported the ultrasonication method for the first time, by applying ultrasonic energy to the solid rubber products such as tires immersed into a liquid. The bulk rubber disintegrates and dissolves subsequently in the liquid and separated. The frequency and intensity of the ultrasonic radiation used was in the range of 20 kHz and 100 W respectively. Even if Pelofsky explained the method of ultrasonication, he failed to describe about the ultimate properties of the devulcanized rubber. Later in 1987, Okuda and Hatano [20] did ultrasonic reclaiming of NR vulcanizate, using 50 kHz ultrasonic energy for about 20 min. This process followed by revulcanization produced reclaimed rubber with similar properties of original rubber. Isayev et al. [21-23] calculated the degree of devulcanization by measuring the cross-link density and gel fraction of the devulcanized rubber by conducting reaction in an ultrasonic reactor. They found a decrease in cross-link density and the gel fraction during the devulcanization process [24, 25] (gel fraction 83 to 64–65% and cross-link density from 0.21 to 0.02 kmol/m^3). They considered the devulcanized sample with a cross-link density lower than 0.06 kmol/m^3 as over treated, and samples with cross-link density higher than 0.10 kmol/m^3 as undertreated and proposed that overtreatment causes main chain breakage and under treatment causes insufficient devulcanization. They made a percolation simulation of the network degradation during ultrasound devulcanization and found an excellent agreement with experimental data. They concluded that devulcanized rubber contained a larger amount of sulfidized molecules which were responsible for cross-linking during revulcanization. They also explained about the ultrasound devulcanization of vulcanized SBR and its revulcanization and found a variation of tensile strength of revulcanized samples from 1.5 to 10.5 MPa and elongation at break from 130 to 250% with decrease in cross-link density of the devulcanizedrubber. They concluded that matrix modification was necessary for the sample size reduction and promoting the ultrasonic reaction and chemical devulcanization.

Kim et al. [26] tried to develop eco-friendly ultrasonic approach by using fewer toxic chemicals. They investigated different ultrasonication methods such as ultrasound only, ultrasound in cosolvents, and ultrasound with oxidant (CAUP-chemically assisted ultrasound process) and gave a general design for the method as shown in the Figure 5. The use of only ultrasound showed relatively lower sulfur concentration than the other methods where synergistic effects take place. The cosolvents can also break more sulfur bonds compared to single solvents. But the higher concentration of sulfur in the solution is found by using ultrasound and oxidant. When H_2O_2 was used as an oxidant, the sulfur concentration increased after 20 min time and it no longer increased after 30 min. The scavenging action of hydroxyl radical can be expressed as

$$H_2O_2 + OH\bullet \rightarrow H_2O + HO_2\bullet$$

These hydroperoxyl radicals, $HO_2\bullet$ have less oxidizing capability than hydroxyl radicals and also after 30 min, all the sulfur from tire molecules will be released. The experiments revealed that the devulcanization technology is successful in obtaining more than 90% devulcanization in tires and finally complete devulcanization by repetition of the process. This method also considered the extended oil recovery, rubber and carbon recovery.

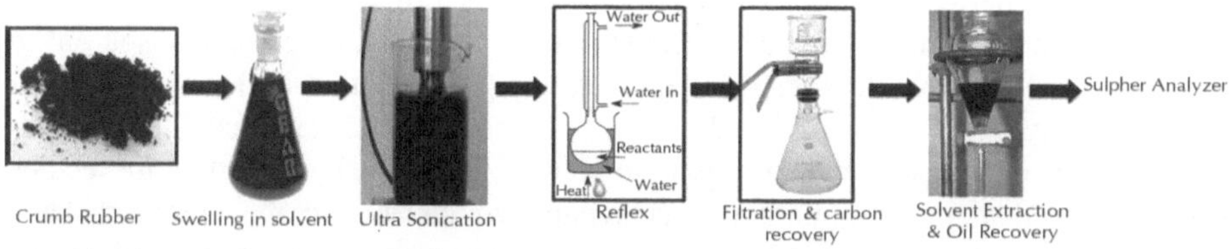

FIGURE 5 Carbon, rubber, and oil recovery, and sulfur removal.

The main problem with ultrasonic devulcanization is that it can cause significant degradation of polymer chains. During ultrasonic treatment, the main chain bond and cross-link bonds breakup as independent random events and the random chain scission results in the formation of soluble branched rubber chains regarded as fragmented gel structure or micro gel [21]. It is found that during ultrasound devulcanization molecular weight of sol fraction decreases and thus it is clear that during irradiation in addition to C–S or S–S bonds, C–C bonds also break. Also, ultrasound is not to be believed as the ultimate removal method because bond scission cannot remove total quantity of sulfur contained in the material.

5.2.1.3 Microwave Irradiation

Microwave radiation can cause devulcanization, which is an essential part in the recycling of rubber. It has been reported that microwave can selectively cleave C-S bonds from C-C bonds and can devulcanize the rubber successfully [27-29]. Thermogravimetry anslysis (TGA) is a powerful tool to analyze rubber devulcanizates obtained after microwave irradiation. Kleps et al. utilized this technique to explore the effect of microwaves on the devulcanization process of Natural Rubber [30]. The TG data can evaluate the degree of thermal destruction of the polymers in devulcanizates by comparison of the thermal parameters and compositions of vulcanizates before and after treatment with microwaves. There is a chance of depolymerization or thermal destruction of the polymer chains during the devulcanization process. The TG data helps to protect the polymers from these unfavorable reactions by establishing the optimal parameters for devulcanization. From the thermograms, the characteristic decomposition temperatures of the vulcanizates and devulcanizates can be obtained. The authors explained the occurrence of intermediates with lower molecular masses than those of the initial elastomers in the rubber devulcanizates due to the degradation of the polymer chains under the influence of MW radiation, by analyzing the TG data.

Microwave heating provides a volumetric heating process at improved heating efficiencies as compared with conventional techniques. In this process, heat is generated volumetrically within the material rather than from an external source. By controllable microwave process, uniform heating within the material can be made possible. The major characteristics of microwave heating are significant waste volume reduction, rapid and selective heating, high temperature capabilities, enhanced chemical reactivity, ability to treat wastes *in situ*, treatment of hazardous components to meet regulatory requirements for storage, ease and remote control, process equipment availability, compactness, cost and maintainability, improved safety, energy savings and so on. It is a cleaner energy source compared with the conventional systems. However, all materials cannot absorb microwaves. Some materials reflect or appear transparent to microwaves and are thus less responsive to heating [31]. Another important thing that to be noted with the microwave irradiation is that this radiation can break the tires rubber molecules into smaller hydrocarbon molecules. And these hydrocarbon molecules when vaporized are useable synthetic gas (syngas) which can operate generators. During this process certain useable by-products like scrap steel, carbon black, oil and so on are also obtained.

5.2.1.4 Gamma Radiation

Gamma ray irradiation is supposed to be a good method for recycling the rubber products. Upon irradiation, the rubber chains break and cause devulcanization. Generally, the mechanical property of a compound is known to decline upon crumb rubber addition. Karaagac et al. succeeded in reducing this rate of deterioration with the use of gamma irradiation. They studied the effects of gamma irradiation on the recycling of used tyre inner tubes by making use of gamma irradiated inner tube wastes and commercial butyl rubber crumbs devulcanized by conventional methods in butyl rubber composites. The rate of reduction in mechanical strength was found to be lower in composites containing irradiated inner tubes than commercial butyl crumbs. The authors observed that there is a better compatibility between the gamma irradiated inner tubes and butyl rubber and so it could be recycled effectively by making composites with butyl based rubber compounds. But this compatibility is found only up to 120 kGy absorbed dose of radiation. Both the virgin butyl rubber and commercially devulcanized butyl rubber crumbs follow this same trend of compatibility [5]. The GTR incorporated thermoplastic matrix possesses mechanical strength similar to thermoplastic elastomer (TPE) and the newly produced composites are also recyclable. But the problem in synthesizing such materials is the compatibility between the rubber particles and thermoplastics. The compatibilization can be achieved by coupling reactions at the interface by using additional reagents such as uncross-linked rubber or by doing surface treatments on GTR particles. Gamma radiation is found to be affecting the [32] *in situ* compatibilization of blends of recycled high density polyethylene (rHDPE) and ground tyre rubber (GTR) powder. In the case of PE, gamma irradiation causes cross-linking under inert atmosphere and oxidation in the presence of air, which lowers cross-linking [33]. But instead, for rubbers, chain scission dominates over cross-linking which enhances in the presence of air [34]. Here in the case of rubber/PE blend system, gamma radiation causes chain scissions within the rubber phase and this is found to reduce the incompatibility with the thermoplastic matrix. There is a chance for the radiation induced cross-linking at the interface of PE and GTR also.

The compatibilization mechanism is explained by three steps molecular chain scission within rubber matrix due to free radicals formation, cross-linking of HDPE matrix and co-cross-linking between the two blend components at the interface. They also observed an improvement in mechanical properties in gamma ray irradiated systems when compared to uncompatibilized rHDPE/GTR blends. The Figure 6 shows the mechanical data for various irradiation doses. This is found to be due to the development of an adhesion between GTR particles and the surrounding thermoplastic matrix. They concluded this as the strong elongation of GTR particles upon deformation of irradiated blends, by *in situ* scanning electron microscopy observations during micro tensile tests.

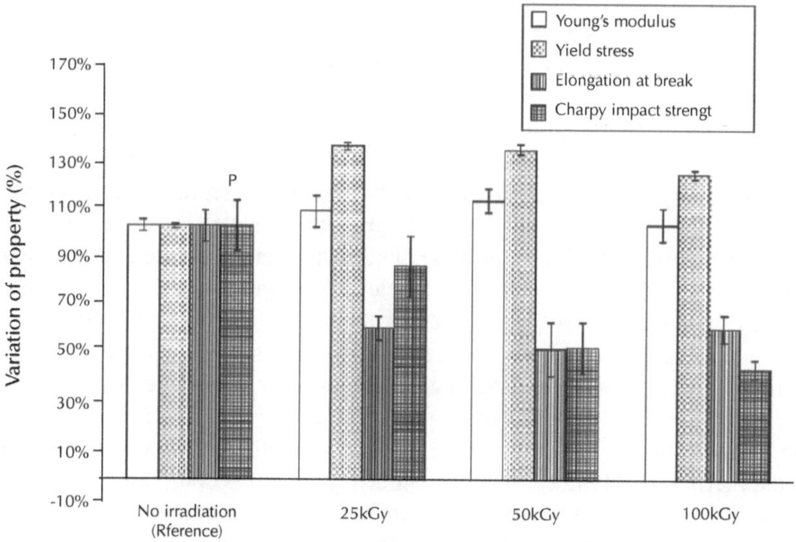

FIGURE 6 Effect of irradiation on the mechanical properties of pure rHDPE. When no irradiation has been applied, pure rHDPE only partially breaks during Charpy impact tests.

It is clear from the figure that the mechanical properties such as elongation at break and Charpy impact strength are significantly increasing for irradiation doses of 25–50 kGy. But the values of Young's modulus show another trend. The radiation induced cross-linking of rHDPE matrix results in the increase of its Young's modulus and yield stress and thus causes the Young's modulus of the blend to decrease slightly. But this cross-linking of the matrix is important only in the case of higher irradiation doses (100 kGy).

5.2.2 Chemical Methods

5.2.2.1 Pyrolysis

Pyrolysis is often referred to an alternative to combustion process as no hazardous emissions are produced and the recovery of solid and liquid material is achieved. Pyrolysis involves the heating of shredded tires in a reactor in an oxygen free atmosphere. The reactor works based on fired fuel, electricity, or microwaves. At first the rubber is softened and there after the chains breakdown to smaller molecules. Later it vaporizes and the collected vapors can be either burned to produce power or condensed to fuel. The uncondensed small molecules remain as gaseous fuel and the minerals separate as solid. From this, steel is removed using magnets and char from the carbon black component remains. Thus the process of pyrolysis of rubber tire material can produce three main components: solid, liquid and gas and their composition is related to the temperature of the thermal treatment. The solid phase constituting approximately 40% weight of the initial sample is mostly carbon black and contains less fraction of mineral matter initially present in the used tire. The gas phase contains a mixture of light hydrocarbons and carbon dioxide, which can be used to provide the energy requirements of the pyrolysis process. Gas chromatography data reports that the tyre pyrolysis gases are composed of CO, CO_2, H_2S, and hydrocarbons such as CH_4, C_2H_4, C_3H_6, C_4H_8 and so on [35]. The liquid phase is a complex hydrocarbon mixture. Both oil and carbon black can be effectively utilized as fuels. The recovery of a carbon black material, thought to be useful as filler, has been reported following the pyrolysis of carbon black filled rubber from tires [36]. The products of pyrolysis can proceed through Diels-Alder type reactions and thus can form cyclic compounds which dehydrogenate to form aromatic systems. It also separates the volatile components from the carbon black, ash, and metal cords that make up the tire. Pyrolysis can also provide H_2 and oil with less sulfur content than present in the original tire. The oil obtained by vacuum pyrolysis of waste rubber tires is used as a coal liquefaction solvent. Coal liquefaction involves two steps. The first step involves heating the coal in the presence of a catalyst, hydrogen gas, and possibly a hydrogen-donating solvent to break down the coal into substances that can be upgraded to a liquid fuel in a second step. The Hydrogen gas should be introduced to react with free radicals that form during the thermal cracking of the coal. Since Hydrogen is comparatively expensive, cheaper source for providing hydrogen is attracting wide attention.

As mentioned above, the waste tire pyrolysis can synthesize hydrogen gas and this can be effectively utilized in the coal liquefaction. Pyrolyzed tire oil hydrotreated with coal converts more than 90% of a bituminous coal to gas, oil and asphaltenes. Over 80% of the coal is converted to products only after 10 min of liquefaction [37]. Mastral et al. [38] analyzed the role of tire rubber in coal tire co-processing when it is added to coal hydrogenation. Here, rubber behaves as a hydrogen donor which could allow the hydrogen pressure to be reduced and therefore to lower the cost of the process. The synergism observed in coal rubber hydro co-processing and the less aromatic nature of the radicals from tire pyrolysis involved in the process helps in improving the nature of the obtained oils. Quek et al. [39] framed a pyrolysis model by formulating heat and mass transfer equations accounting the different extents of thermal lag as the tire is heated at different heating rates. The model predictions are found to be in good agreement with experimental data especially between the temperatures from 450K to 1000K. They established that the incorporation of heat and mass transfer processes in a pyrolytic model can improve the quantitative understanding of tire pyrolysis. However the conditions of pyrolysis may vary greatly in literature. This is because the pyrolysis yields and characteristics of the products obtained depend not only on the feedstock and operating conditions used for the experiments, but also on the specific characteristics of the system used, such as the size and type of reactor, the efficiency of heat transfer and the residence time. Based on this fact, Laresgoiti et al. studied the nature of pyrolysis liquids at 300, 400, 500, 600, and 700°C. They concluded from the distillation data that the tire oil derived fractions could be used both as automotive diesel oil and heating diesel oil. Since the tire oil does not fulfill the requirements of commercial diesel oils, a series of hydro treatments and/or blending them with petroleum refinery streams is required, in order to use them as diesel oils [40]. The pyrolysis is a clean process without any emission of waste if it is properly done. The process can be either batch or continuous. Catalysts can also be used to accelerate the decomposition.

5.2.2.2 *Super Critical Extraction*

Because of the presence of chemical cross-links in tires, they cannot be dissolved in usual solvents. Super critical fluids can cause thermolysis of scrap tires and can dissolve the rubber particles in the liquid. This technique has several advantages than the conventional thermolysis such as no discharge of toxic gas, low temperature operation and fast decomposition time and recovery of hydrocarbon. During this practice, the intra particle mass transfer processes become fast and thermal decomposition acts as the controlling step, while in pyrolysis, the overall reaction is controlled by thermal decomposition only especially with powdered solid. Supercritical fluids such as water, methanol, cyclohexane, n-pentane, toluene, tetrahydrofuran, and so on are used for thermolysis of tires at low temperature. Super critical water is almost as effective as super critical pentane in the tire extraction process [41]. Chen et al. used supercritical H_2O and CO_2 to controllably depolymerize tire and natural rubber [42]. According to them, after supercritical extraction, the tire samples can generate various products as described in Figure 7(a) gaseous part, which is not collected, an aqueous phase (if the supercritical fluid used is water), an organic phase (existing as a free layer or absorbed on the carbon black), and a solid phase, which is carbon black. The experiments showed that that supercritical H_2O yields a higher degree of decomposition of tire rubber than supercritical CO_2. This is attributed to the high dielectric constant of supercritical water, which would support heterolytic reactions and/or the direct attack of water as a nucleophile.

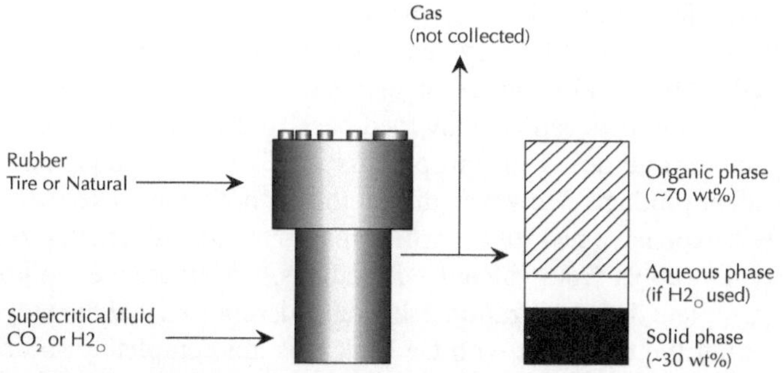

FIGURE 7 Schematic representation of materials generated from supercritical fluid depolymerization of tire rubber.

Usually the chemical reactions like pyrolysis proceeded *via* free radical mechanism and hydrolysis through nucleophilic attack yield different product states since they result in homolytic as well as heterolytic pathways. The ionic properties of the supercritical fluids can be controlled by temperature and pressure and thereby offers a control over the reaction and thus results in product selectivity. The recycling can be made more effective by coupling extrusion and supercritical fluid method. For this, Tzoganakis et al. injected supercritical CO_2 in to a twin screw extruder to swell the rubber crumb and to facilitate the otherwise impossible rubber extrusion process. [43]. The feeding flow rate of rubber to the extruder, the screw configuration, the screw type and processing temperature are found to be the main factors affecting the extrusion process and thereby the process ability. The concentration of super critical fluid CO_2 is found to have little influence on the devulcanization. The super critical fluid extraction can be combined with gas chromatography and thus the decomposition percentage is calculated by Bhatti et al. using the equation

$$\text{Decomposition \%} = 1 - W_F/W_I \times 100 + W_C\%$$

where W_F denotes weight of residual solid after drying, W_I the initial weight of sample tire and W_C the weight % of carbon black and inorganic materials in the sample tire. [44]. They obtained highest decomposition of 94.2% at a temperature of 308°C and at a pressure of 750 psi. The supercritical extraction process depends more on the temperature than the pressure.

5.2.2.3 De-link process

De-link R is a most recent method for the recycling of vulcanized rubber. This is introduced by two polymer scientists, Dr B. C. Shekhar and Professor V. A. Komer [45]. It is a mechanochemical process which involves the addition of a reactant called De Link, a patented devulcanising agent to rubber product. The De-link R exists as a rubber master batch form with natural or synthetic rubbers rubber as binder. It comprises mercaptans, thiazoleor dithiocarbamates and other simple peptizers used for rubber [46]. This process is supposed to be the most useful method for rubber recycling not only because of its economic and environmental friendly nature but also its ability to recycle both natural and synthetic rubbers containing sulfur/metal oxide vulcanizing agents. When De Link R is added to rubber products in a mechanical mixer, it functions along with the shear force to break the sulfur cross-links present inside the rubber and results in the devulcanization. The recycled product obtained has the same properties like unvulcanized rubbers and it can be revulcanized again. Using this process, upto 30% of the recycled products can be converted back to the original materials without much affecting their properties. Thus this process is effective in producing raw rubber from the used rubber products. However, this method is not widely used because of the availability of less expensive raw rubber which makes the manufacturers to use it than making eco-friendly rubber from vulcanized products. Also there are reports showing the inability of complete devulcanization. Ishiaku et al. reported that the recycled compound obtained after the treatment with De Link R is not completely soluble in suitable solvents and thus this process did not cause complete devulcanization [47]. This aspect is further proved by the cure time and tensile strength studies. They observed an increase in tensile properties with De Link R amount upto an optimum value of 6 phr. This is contradictory to the observation made by Kim et al. in 2004 where they analysed the role of De Link R in determining the devulcanisation of natural rubber and found better properties at the optimum amount of the reactant 20 phr.

5.2.3 Biological Method

Devulcanization or removal of sulfur bonds can be done with help of microorganisms as well. But since it is hard to control the circumstances with microbial agents, this method is not popular as other conventional processes discussed so far. Anyway a lot of studies are going on in this area as it offers the most environmental friendly method. Many mesophilic and thermophilic microorganisms existing can oxidize sulfide minerals and reduced valence sulfur compounds. These microorganisms are chemolithoautotrophs that can derive energy from the oxidation of the above substrates and their carbon for cell material production from CO_2. Examples of mesophilic bacteria include *Thiobacillusferrooxidans*, *T. thiooxidans*, *Leptospirillumferrooxidans*, *T. organoparus*, *T. thioparus*, and so on. Thermophilic bacteria such as *Sulfobacillusacidocaldarius*, *Sulfobacillusthermosulfidooxidans* are also effective. During microbial devulcanization method the rubber molecules should be the substrate for microbial growth. Kim and Park reported [48] devulcanization of rubber using the microbium, *T. peromatabolis*. They unvulcanised crumb rubber with the help of microbium and then compounded the unvulcanised product with NR. Cross-link density calculation from swelling experiments was mainly used to find out the degree of unvulcanization. The authors also compared the results obtained with chemical treatment and found that, for more than 5 phr crumb rubber compounds, the microbial treatment is more effective than chemical treatment in maintaining tensile and elongation properties. The sulfur removal from rubber tire by bacteria can be explained by the following steps. Sulfur oxidizing micro-organisms had oxidized the sulfur present in the rubber to sulfate, that is sulfuric acid.

Inorganic portion: $S + 1.5\ O_2 + H_2O \rightarrow H_2SO_4$ (in presence of bacteria)
Organic Portion: $R\text{-}S_x\text{-}R + y\ 1.5\ O_2 + y\ H_2O \rightarrow y\ H_2SO_4 + R\text{-}S_{x\text{-}y}\text{-}R$ OR
$R\text{-}S_x\text{-}R + x\ 1.5\ O_2 + x\ H_2O \rightarrow x\ H_2SO_4 + R\text{-}R$ (both steps happens in the presence of bacteria).

Bredberg et al. used anaerobic sulfur reducing archae on *Pyrococcusfuriosus* for the reduction of sulfur in rubber [49]. The anaerobic process offers greater specificity as it does not involves hydrocarbon oxidation. The study reveals an increase in sulfur production when the microorganism was grown in vulcanized natural rubber, indicating their ability to utilize sulfur. They also investigated the tensile strength, stress relaxation, and swelling measurements and found that the mechanical properties are affected by microbial treatment. It is clear from the high stress at break values of microbial treated crumb rubber that, this treatment is as effective as oxidative treatment. They concluded that the microbial treatment can improve the properties of rubber materials to a greater extent.

Breaking of sulfide bonds on the surface is expected to make the polymer chains more flexible, and thus yielding more unreacted unsaturated bonds accessible for revulcanization. Christiansson et al. carried out surface devulcanisation studies with microorganisms *Thiobacillusferrooxidans, T. thioparus, Acidianusbrierleyi* and an archeal isolate TH2 Lund [50]. They examined the sulfur content after 23 days of treatment and found that *Thiobacillusferrooxidans* was the most efficient in terms of released sulfate among all the other organisms used. Since all the sulfur molecules added to rubber are not utilized for vulcanization, traces of sulfur can be present as adsorbed on the surface and sometimes can form vicinal cross-links. The authors suggested a suitable method for removing these particles even though no investigations on the origin of the extracted sulfur have been made. They also did not mention about the processing conditions that can influence the desulfuration process. Surface treatment of GTR with *S. acidocaldarius* to provide an improved surface chemistry for compounding into virgin rubber materials was reported by Robert et al. [51]. The surface treated GTR particles are then compounded with various rubber systems and the physical properties of these base rubber compounds got improved.

Absence of harmful or toxic chemicals is considered to be the main advantage of this method. This method is normally not energy intensive. Also the specificity of microorganisms and enzymes results in less unwanted degradation of the material. However, the method is not free from demerits. The higher sensitivity of microorganisms towards many chemical substances, including rubber additives is considered to be the major drawback. This problem is solved by leaching of tire materials with organic solvents before treating with microorganisms in order to remove the additives such as antioxidants, accelerators, and so on in rubber. But, this is not highly appreciable since it involves the use of organic solvents. Even though, studies show the removal of sulfur bonds from rubber by the microorganisms, advanced analytical techniques (SEM-EDS, ESCA) reveals that the removal is only a surface phenomenon. So, effective studies are to be carried out in this area to minimize all the drawbacks and to establish the most environmental friendly method.

5.3 ADVANTAGES AND APPLICATION OF USING RECLAIMED RUBBER

As explained, the recycled rubber can be effectively utilized for a number of applications. It has several uses, of which the most important is reinforcing the other systems. Recycled rubber is proved to be good for reinforcing concrete as well as several other polymers. The important applications are explained in detail in this section.

5.3.1 Blending with Concrete/Cement

This method is the most common and widely used method for rubber recycling. For better applications of concrete, the low unit weight, high strength, toughness and impact resistance are necessary. It is found that the use of waste automobile tire in the form of chips, scraps or fibers can impart all these qualities to concrete. Rubberized concrete mixtures possess lower density, increased toughness and ductility, higher impact resistance, lower compressive and splitting tensile strengths, high resilience and more efficient sound insulation. It also shows high plastic energy capacity and thus exhibiting high strains, especially under impact effects. Different kinds of tyres have been employed as partial substitute of natural aggregates in concrete: scrap tyres obtained by simple grinding and without further purifications thus including steel and textile fibers in their composition [52, 53], crumb rubber obtained by cryogenic process [52], milled tyre rubbers treated with sodium hydroxide solution to achieve a better adhesion with the cement paste [54], scrap truck tyre rubber [55], tyres tread [56], and so on. However, regardless of the type of tires a decrease in concrete compressive strength in the mixture was always detected with the increasing amount of rubber phase. Since, the rubbers absorb more energy the plastic deformation at the time of fracture increases by the rubber replacement in concrete [57]. Lee et al. [58] investigated the flexure and impact strength of Styrene–Butadiene–Rubber (SBR) latex modified crumb rubber filled concrete and observed a high flexure and impact strength for concrete compared to the conventional Portland cement and latex modified concretes. The reason was explained as the formation of a thin film of SBR latex at the interface of cement mortar and aggregates, increasing the interfacial bonding strength. But the high cost of SBR latex limited its applicability. Topcu et al. also reported that the addition of rubber aggregate into concrete increase concrete's impact resistance, which became more prominent in the concrete samples containing larger rubber aggregates. This is attributed to the fiber structure of aggregated rubber, which gives the concrete more flexibility and capacity to take in strokes [59]. Eldin and Senouci [52] investigated the strength and toughness of concrete with a portion of aggregates replaced by waste tire chips and observed that the compressive strength and split tensile strength were reduced, while its toughness and ability to absorb fracture energy were enhanced significantly. Khatibet al. [53] used fine crumb rubber and tire chips to replace a portion of fine or coarse aggregates and they also found a reduction in strength for rubber filled concrete, even if its toughness was enhanced. From these studies it is clear that waste tire rubber modified concrete has high toughness whereas low strength and stiffness. Among, the various steps taken to improve the strength and stiffness of waste tire modified concrete, two ways are of particular interest. One is the surface treatment of waste tire powders and the other is using thin waste tire fibers. Pretreatments may vary from washing rubber particles with water to acid etching, plasma pretreatment, and various coupling agents. This enhances the strength of concrete containing rubber particles through a microscopic increase in the surface texture of the rubber particles. Segre and Joekes [54] found that NaOH surface treatment on waste tire chips can increase rubber/cement paste interfacial bonding strength and thus can improve both the strength and toughness. They noticed that higher the NaOH concentration, better the adhesion.

Short fibers like steel fibers can increase the mechanical strength, toughness, and cracking resistance of concrete to a great extent, there is a possibility of improved mechanical properties for fiber shaped waste rubber incorporated concrete than all the other forms of reinforcement. As a result, rubber fiber modified concrete may be accepted and used in load carrying structures. Since, macroscopic rubber fibers or rubber straps can be produced using mechanical cutting, the cost for cutting rubber fibers is also much lower than that for producing crumb rubbers. Li et al. studied about the scrap tires wastes in the form of fibers produced in the United States and found that highway construction provides a significant market potential for waste tire recycling. The dry process of synthesis and the low cost of waste tire modified concrete make it superior over the waste tire modified asphalt produced by wet process. In addition, the resulting concrete has a very high toughness, higher cracking and fracture resistance which is desirable because conventional concrete is a brittle material. They evaluated the feasibility and performance of waste tire modified concrete using larger sized fibers and NaOH treated chips and also the factors such as physical anchorage, fiber aspect ratio, tire resources, and hybrid fiber reinforcement that may affect the performance of modified concrete [4, 60]. They found that the rubber fibers are better than waste rubber chips. Guneyisis et al. observed the influence of silica fume on the properties of the rubberized concrete by using crumb rubber and tire chips. Silica fumes can increase the homogeneity and decrease the number of large pores in cement paste [61], both of which would lead to a higher strength material. The use of silica fume resulted in a denser interface between cement paste and coarse aggregates [62]. Self-compacting concrete (SCC), is a new type of concrete that attains higher compressive strength and durability in comparison with ordinary Portland cement concrete (OPCC). In spite of the presence of fine filler promoting the formation of very compact microstructure and allowing high values for compressive strength, the failure behavior in SCC is still brittle. Bignozzi et al. explored the possibility to design self-compacting rubberized concrete (SCRC) by joining the characteristics of SCC (high flow ability, high mechanical strength, low porosity, etc.) with the tough behavior of the rubber phase and they succeeded in obtaining a building material with more versatile performances [63].

Most recently, Ling explained the mechanism of rubber concrete composites [64] as given in the Figure 8.They noted that the compressive strength of plant made RCPB increased, when the proportion of crumb rubber in concrete was 10%. Upon the addition of a small proportion (10%) of crumb rubber to the mixture, the soft rubber particles are easily distorted and filled the voids between the solid particles (natural aggregates) under a compression force of plant made machine. This filling mechanism reduces the porosity by filling up the free pore volume in the concrete mixtures (Figure 8(b)) and also at this situation, the rubber particles had bonded well with cement matrix (Figure 8(a)) which in turn resulted in a better compressive strength of RCPB. A perfect adherence between rubber and cement matrix has also been reported [65].

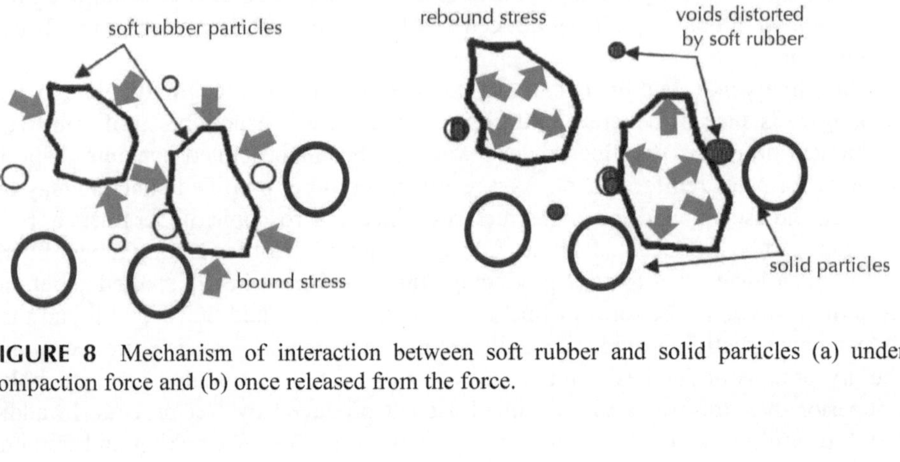

FIGURE 8 Mechanism of interaction between soft rubber and solid particles (a) under compaction force and (b) once released from the force.

Rubber wastes also include large quantities of other rubber based powdery forms, obtained from mechanical shredding of rubber derived from automobile industry waste. No organized collection system has been set up to handle these waste products, which are often simply discarded at dumpsites. This causes significant environmental, health, and aesthetic problems. Benazzouk et al. successfully used these rubber waste particles, as a raw material, to develop lightweight construction materials. They investigated the effect of rubber particles addition on the physico-mechanical and water absorption properties of the composites [66]. The composites were manufactured by adding rubber particles in cementitious matrix and the dry unit weight, elasticity, dynamic modulus, compressive, and flexural strengths, strain capacity, and water absorption were found to be improved. The effects of epoxidized natural rubber (ENR-50) as a compatibilizer on the properties of styrene butadiene rubber/recycled acrylonitrile-butadiene rubber (SBR/NBRr) blends were also reported in the literature [67]. This report considered the state and federal regulations for disposal of scrap tires and the toxicity of tire rubber to humans and the environment since the disposal of scrap tires in landfills has proven to be a problem as the whole tires can "float" to the surface and break the cap of the landfill.

The performance of asphalt concrete can be enhanced by the addition of tire rubber, which can provide a satisfactory level of service for the expected traffic [68].The manufacturing of modified asphalt is a complex process as it shows different mechanical properties with every level of tire rubber gradation, mixing temperature, aggregate gradation, tire rubber ratio, binder ratio, compaction temperature, and mixing time. Tortum et al. determined the optimum working conditions such as asphalt concrete's Marshall Stability, flow, unit weight, void, and binder-filled void, using Taguchi method [69].This method proves the ease and flexibility to design complex materials like asphalt concrete. The mechanical and structural properties of asphalt concrete, modified tire rubber were investigated using Marshall Test Machine. The Taguchi method consists of experimental planning with the objective of acquiring data in a controlled way to obtain information about the behavior of a given process. The advantages of Taguchi method on the conventional experimental design methods are maintaining the experimental cost at a minimum level, minimizing the variability around the target when bringing the performance value to the target value and less time.

5.3.2 Blending with other Polymers

The TPEs are flexible, low modulus polymers used in a range of applications, from automotive interiors and heat resistant tubing to "soft touch" grips for many consumer goods. The TPEs are generally considered to include olefin TPEs (TPOs), thermoplastic vulcanizates (TPVs), styrenic block copolymers (SBCs), thermoplastic polyurethanes (TPUs), polyether block amides (PEBA) and copolyesters (COPE).The TPE based on rubber–plastic blends can be divided into two main classes, thermoplastic olefin (TPO) and thermoplastic vulcanizate (TPV). Since the dispersed rubber phase is not cross-linked in TPO, it can be easily prepared at relatively low cost. But, the preparation of TPV requires a more complicated process as the dispersed rubber phase must be cross-linked during mixing, mostly through dynamic vulcanization or an *in situ* cross-linking process. Cross-linking in TPVs enhances properties such as tensile strength, elastic response, and fluid resistance compared to uncross-linked TPEs. In order to get good mechanical properties for TPEs, combination of high shear stress during mixing, complete homogenization of the dispersed rubber particles and avoidance of local overheating and polymer degradation are necessary. Rubbers such as ethylene propylene diene monomer (EPDM) and ethylene propylene monomer (EPM) are generally selected for TPE preparation to avoid excessive degradation of the rubber phase. The TPE prepared from natural rubber, thermoplastic natural rubber (TPNR), is also important. The TPNRs can be prepared from the combinations of natural rubber with various conventional plastics, such as polypropylene (PP) [70], low-density polyethylene (LDPE) [71], linear low-density polyethylene (LLDPE) [72] as well as high density polyethylene (HDPE) [73, 74] by melt mixing technique. Pongdhorn et al. prepared TPNR without dynamic vulcanization, by blending HDPE with pre-vulcanized natural rubber powder (NRP) and their properties including recyclability were investigated [75]. Polypropylene (PP), due to its low cost, process ability and good balance of properties is also used to make TPEs [76]. Awang et al. investigated the effects of partial replacement of waste tyre dust (WTD) on PP and the properties of PP/WTD blends as well as the influence of trans-polyethylene rubber (TOR) and dynamic vulcanization on the PP/WTD blend properties [77]. Also they prepared thermoplastic blends based on PP by using waste rubber from scrap tyres and a renewable material source such as natural rubber latex. Besides, consuming less energy when produced, natural rubber latex also possess adhesive properties, low viscosity and low surface tension which render some improvements to the interaction between PP and WTD [78]. Besides recyclable PP/WTD blends, the use of WTD in the blends promotes recycling of waste rubber from scrap tyres.

The reinforcing effect of rubber in various polymers is generalized in to Table 1. The advantages of rubber incorporation in polymers as well as the enhancement in various properties can be clearly understood from this table.

TABLE 1

S. No.	Composition			Fracture toughness	Tensile strength (Mpa)	Modulus	Elongation at Break (%)
1	Epoxy			0.98	–	–	
[79]	Epoxy	GTR (5vol %)	VS(2wt%)	1.13	–	–	–
			AS(2wt%)	1.43	–	–	–
			AA(5wt%)	1.26	–	–	–
			AA/BP(5wt%)	1.54		–	–
2 [67]	SBR(95phr)	NBRr(5phr)	ENR(0phr)	–	23.5	2.7Mpa (M100%)	520
			ENR(10phr)	–	24.2	2.9Mpa (M100%)	510
	SBR(50phr)	NBRr(50phr)	ENR(0phr)	–	10.5	4.1Mpa (M100%)	375
			ENR(10phr)	–	11	4.3Mpa (M100%)	250

TABLE 1

S. No.	Composition			Fracture toughness	Tensile strength (Mpa)	Modulus	Elongation at Break (%)
3	rHDPE	–	–	–	–	100%	100
[32]		GTR (10wt %)	–	–	–	85%	80
		GTR (70wt %)	–	–	–	22%	55
4 [80]	LDPE	–	–	–	8.2	400Mpa	175
	LDPE (70wt %)	GTR (30wt %)	–	–	7	340Mpa	70
	LDPE (50wt %)	GTR (30wt %)	EVA (20wt %)	–	5.9	240Mpa	150
5	RPE (75wt %)	GTR (25wt %)	–	–	6.2	82Mpa	94
[81]	RPE (50wt %)	GTR (50wt %)	–	–	4.3	48Mpa	70
	RPE (25wt %)	GTR (75wt %)	–	–	3.2	18Mpa	106
6 [82]	WPE	–	–	–	15	9.8Mpa (M100%)	490
		WRP(5phr)	–	–	13	10.2Mpa (M100%)	400
		WRP(25phr)	–	–	9	9Mpa (M100%)	250
7	PP (80wt %)	WTD (20wt %)	–	–	17	710Mpa	6
[78]	PP (80wt %)	WTD (20wt %) (modified with latex)	–	–	21	860Mpa	14
8 [83]	SBR(85phr)	NBRr(15phr)	CB/sil (10/40phr)	–	9	2 Mpa(M100)	700
	SBR(85phr)	NBRr(15phr)	CB/sil/Si69 (10/40/3phr)	–	14	2.4 Mpa(M100)	750
9 [84]	NR (70phr)	RRP (30phr)	–	–	22	1.2 Mpa(M100)	710
	NR (40phr)	RRP(60phr)	–	–	15	1.4 Mpa(M100)	500
10 [85]	A–HDPE(40phr)	GTR(30phr)	EPDM(30phr)	–	4.5	4.5 Mpa(M100)	133
	A-HDPE(40phr)	–	EPDM(60phr)	–	3.5	3.2 Mpa(M100)	200
	A HDPE(100phr)	– –	–	–	27	203 Mpa(M100)	–
11 [86]	PP(100phr)	–	–	–	37	1150MPa	10
	PP(40phr)	RR(60phr)	–	–	10	450 MPa	44
	PP(40phr)	–	NR(60phr)	–	6	400 MPa	105
12	SBR(100phr)		–	–	92(%)	–	50
[87]		SBRr(50phr)	–	–	112(%)	–	50
		SBRr(80phr)	–	–	97(%)	–	55

TABLE 1

S. No.	Composition			Fracture toughness	Tensile strength (Mpa)	Modulus	Elongation at Break (%)
13 [14]	SBR(100phr)	–	–	–	2.335	1.25 Mpa(M100)	432
	SBR(80phr)	RR(20phr)	–	–	2.781	1.59	377
	SBR(80phr)	RR(20phr)	Spindle oil(4phr)	–	12.577	4.196	509
	SBR(80phr)	RR(20phr)	–	–	5.017	2.28	445
14 [88]	LDPE (13.1wt %)	–	APP/EVA/rPU (25/35.9/25)		2.3	82.3	6
	LDPE (9.1wt%)	RR (15wt %)	APP/EVA/rPU (25/34.9/25)		1.6	70.6	32
15 [89]	BR	–	–	–	15	–	540
		Crumb rubber (size of particle 0–0.4mm)	–	–	7.5	–	470
		Crumb rubber (size of particle 0.4–1mm)	–	–	10.5	–	325
16 [90]	–	EPGRT(100phr)	–	–	7.8	6.5(M100)	200
	NR(60phr)	EPGRT(40phr)	–	–	16	0.7(M100)	830

VS—Vinyltriethoxysilane; AS-3—Aminopropyltriethoxysilane; AA—Acrylic Acid; GTR—Ground Tyre Rubber; SBR—Styrene butadiene rubber; SBRr—recycled Styrene butadiene rubber; NBRr—Recycled acrylonitrile-butadiene rubber; ENR—Epoxidized natural rubber; rHDPE—recycled High density polyethylene; LDPE—Low density polyethylene ; RPE—recycled polyethylene; WPE—Waste polyethylene; WRP—waste rubber powder; WTD—waste tyre dust ;PP—Poly Propylene; Si69-bis-3-(triethoxysilyl)-propyl)-tetrasulphide ; Sil-Silica ; NR—Natural Rubber ; RRP—Recycled rubber waste; A-HDPE—Acrylic modified High Density Poly Ethylene ; EPDM—ethylene propylene diene monomer; RR—Recycled Rubber ;APP—Ammonium polyphosphate ;rPU—Recycled polyurethane foam ; BR—Basic rubber ; EPGRT—Extrusion processed ground rubber tire.

The thermodynamically incompatibility is one of the main problems with blending of polymers. In most cases, even between closely related polymers, the compatibility is on low level, which leads to phase separation and weak adhesion between the phases. In rubber recycling process, the GTR can be blended with thermoplastic polymer matrices, but they have limited mechanical performance due to the compatibility problem. Among the thermoplastic polymers, the recycling is easier for poly-ethylene because of the relatively simple build-up of the PE molecules. Since PE molecules are non-polar, the addition of reactive agents like peroxides, anhydrides and so on [81, 91-94] or polar ethylene vinyl acetate copolymer (EVA) can improve the compatibility of the two phases and thus better interaction with the GTR. High energy irradiation treatment can produces free radicals on the surface of the GTR particles, which attack the PE to form covalent bonds between the two phases and also can cause the cross-linking of PE resulting in compatibilization. [95, 96]. Meszaros et al. used a combined compatibilizing influence of copolymer EVA and high energy electron beam to study the blending between PE and GTR. The samples after melt-mixing were treated by high energy electron beam and they found enhanced properties for the obtained composites [80].The effects of different sizes of recycled acrylonitrile-butadiene rubber (NBRr) and blend ratios on curing characteristics, mechanical properties and morphological properties of styrene butadiene rubber/recycled acrylonitrile-butadiene rubber (SBR/NBRr) blends was reported by Noriman et al. [97]. They investigated the effects of epoxidized natural rubber as a compatibilizer on the properties of Styrene butadiene rubber/recycled NBR blend system as well. The effects on cure characteristics, mechanical properties, FTIR, DSC analysis, and also morphological behavior of SBR/NBRr were studied [67] and found that the properties were significantly enhanced. Matko et al. prepared flame retardant material by blending recycled rubber tyres, low density PE, EVA copolymer and an intumescent additive system consisting of waste polyurethane foam and ammonium polyphosphate [88]. They observed that the rubber powder containing compound considerably reduced CO_2 and CO emission compared to reference material. Polycarbonate/recycled rubber blend system is another important area of study. Zribi et al. [98] added carbon black particles to polycarbonate/waste tire rubber blend to prepare conductive composites and studied the percolation effect. They found a positive synergy between the carbon black and crushed rubber particles. The synthesized composites had excellent thermoelectrical and chemoelectrical properties making them useful in heating and sensing applications. Eco-friendly plastics can also be obtained by combining the elastomeric crushed rubber particles and thermoplastic polycarbonate [98]. Blending rubber with the matrix cause cavitation effect, this leads to a plastic shear yielding effect in the surrounding matrix. The suitable formulations for the plastics possessing good mechanical properties are determined. Bourmaud et al. used nanoindentation technique to measure the mechanical properties of recycled polycarbonate/rubber blends [99]. Using nanoindentation technique, the hardness or modulus at various parts of a sample can be determined. The crushed tire reclaimed rubber particles are modified using flame treatment and using a solvent, chloroform before adding to the polycarbonate. It is observed that the rubber particles act as plasticizer. The surface treatment increases the hardness and modulus value to a great extent compared to the unmodified rubber particles. The nanoindentation tests on the blend also showed the presence of an interphase region. Later in their work, they developed conductive polymer composites from recycled rubber/polycarbonate system [100]. The abrasive wear behavior and kinetics of reaction are studied based on the surface treatment done to the rubber particles. The obtained composite is useful as smart materials for sensing applications. The deterioration of mechanical properties due to the poor interfacial interaction between PC and rubber particles is minimized by the surface treatment of rubber particles, especially using methanol. The wear resistance of this system is reported to be the highest.

5.3.3 Rubber Based Wood Materials

The growing demand for wood based panels has led to continuous efforts to find new resources as an alternative to wood. In addition to this problem, the demerits of wood based structures should also be addressed. The lignocelluloses cell wall of wood materials can absorb or lose moisture from or to the atmosphere and can swell or shrink. Due to this the stability of wood based products are less when there is a change in the climatic conditions. Value-added wood based panels manufactured from recycled materials are considered to be a solution to all these problems. The waste tire rubber is an exceptional raw material for the wood composite panel because of its excellent energy absorption, better sound insulation, durability and abrasion resistance, anti-caustic and anti-rot [101] nature. The tire rubber is almost hydrophobic and is only minimally affected by atmospheric humidity. Thus, the incorporation of tire rubber in wood improves the lifetime and usability of rubber products and decreases the hazards due to waste rubber tires at the same time. Reports show that waste tires can be effectively recycled by manufacturing wood rubber based composites [102]. Song and Hwang [103] examined the wood/waste tire composites and found that the ratio of wood fiber to rubber particles and amount of a surfactant, diphenylmethanediisocyanate (MDI) were the significant factors that influenced board mechanical properties. Oriented Strand Board (OSB) is a structural panel made of wood strands that are arranged in cross-oriented layers, similar to plywood. It has similar strength and rigidity, panel size and thickness, fastener performance and paint ability as the other structural panels. Such OSB panels strengthened by waste tire rubber chips were manufactured by Ayrilmis et al. [104]. For the manufacturing, melamine/urea formaldehyde and diisocyanate were used as adhesives. They investigated the water resistance and mechanical properties such as modulus of rupture and modulus of elasticity along with the internal bond strength. Waste rubber improved the water resistance and as the amount increases, it reduces the hydroxyl groups present in the OSB panels and thus decreases the swelling thickness. The adhesive resin improved the binding strength significantly.

5.4 ENVIRONMENTAL HAZARDS

5.4.1 Problems before Recycling

As already discussed, in the introduction part waste rubber is creating a lot of problems to the environment. Because of the large volumes and about 75% void space, tires consumes much space and thus creates sever space problem, if discarded. The abundant and indiscriminate disposal of scrap tires can cause serious environmental as well as health problems. When the tires are piled, it can become the habitat of rodents, reptiles and also the water trapped in the tire makes the place for breeding mosquitos, which leads to severe health problems. The accumulated tire can also cause fire hazards. Tyre fire by products can pollute the surface as well as sub surface water and soil. Inhalation of components of tire rubber upon burning or dust particles from tire rubber can be irritating to the respiratory system and can cause asthma and sometimes even mutagenic or carcinogenic effects. A few adverse effects are shown in Figure 9. Sometimes, the dumping of tire can block the drainage system. The discarded tire stockpiles can also cause wasting of land suitable for agriculture and will reduce the soil fertility as well. Tires can trap gases like methane inside it. Due to this tires become buoyant, and can eventually damage landfill liners.

FIGURE 9 Environmental problems due to rubber accumulation.
http://www.constructionweekonline.com/article-8947-in-video-beeah-rubber-recycling-plant-in-action/
http://discardstudies.files.wordpress.com/2011/11/kirby-solar-art0-g50c6tkf-1tire-fire22-jpg.jpg

There is a chance for leaching of various ingredients such as additives, stabilizers etc which are compounded with rubber during tire manufacture. The process of leaching varies according to the pH and other conditions of soil and water and it will be quiet high in the case of shredded tire pieces due to their large surface area. Leaching causes these additives to move out to the environment and thus killing the useful microorganisms in the soil. Complete burial of tires inside the soil is another disposal method, but this results in the trapping of air inside it and finally it moves out to the surface. This causes instability of the sites as well. The incineration and thermal decomposition of tires can generate toxic gases which cause air pollution. Even if the tire recycling is a solution to all these problems, due to its less cost effectiveness and environmental problems, most of the producers dump the waste rubber products legally or illegally. The disposal of tires to overseas as well as burning tires for energy is also not environmentally friendly. Thus, the waste tire stream causes environmental, human health and aesthetic problems significantly.

5.4.2 Problems due to Recycling

Even though the recycling of rubber is a boon when thinking about the problems due to dumping of tires and other rubber products in soil, it imparts serious complications also. The recycling is considered to be unsustainable because of the high cost involved. The expense is high when compared to the price of raw rubber. The smoke and particles discharged to the atmosphere from the rubber recycling factories can cause severe health problems to the workers, employees and residents in the nearby areas. Workers safety is an important problem observed while recycling. The incineration of rubber can produce sooty flame and can also releases Zinc Oxide to the atmosphere. Unless, it is carefully controlled and collected, it can become a big problem to the environment. The incineration of synthetic rubber like chloroprene can produce HCl vapors as it contains chlorine atoms in them. High temperature can also affect the furnace as the highly combustible material can explode inside. This is an exceptional case particularly when the operating person is not aware of the facts or with improper management [105]. All tires cannot be recycled completely due to several reasons. Only a few percentages of tire wastes are converted to useful products. This will also cause issues to the nature. The recycling can cause the formation of tire oil 45%, carbon black 35% steel wire 10–12% and waste gas 8–10%.The carbon content recovery from this material is really difficult. The effective disposal of solid waste after pyrolysis is another problem associated with recycling. The recycling process involves the use of chemicals and also it consumes more energy. Unless using sophisticated techniques to control the temperature, feeding of tyres inside the equipment and to keep the emission of toxic gases within the environmental limits, techniques such as pyrolysis become the most serious hazard to the nature. By following a proper manufacturing technique and by keeping the quality of delivery, the wastes at production can be reduced.

5.5 APPLICATIONS

The recyclability and re usability of rubber are discussed so far. Generally the recycled rubber can be used in various fields such as landscaping, horticulture, gardening, sports, construction, transport, and so on. About 10% of the scrap tires generated are used as playground and other sports surfaces, and in rubber modified asphalt. Crumb rubber can be used on running and jogging tracks, athletic fields and golf courses because of its resiliency and durability. Shredded tires have been shown to effectively remove organic compounds from leachate in landfills and elsewhere, and ground rubber also has been found to adsorb metal contaminants. Shredded or crumbed rubber can also be used for surfacing for equestrian, sports and safety surfaces, carpet underlay, street furniture and acoustic barriers, as well as to incorporate into new tyres manufacturing. Ground rubber powder is used as filling material in a wide range of applications, together with different matrices such as new rubber material, plastics and cement. Shoe soles, plant pots and swings for children's playgrounds are other possible application. The major fields of applications of crumb rubber are summarized in Figure 10.

The process ability of compounds containing recycled materials is excellent and applicable as fire retarded thermoplastic rubbery material for engineering purposes such as floor covering or interior panel in construction of buildings [88]. The reuse of rubber products from used tires has the potential to make a substantial contribution to reducing carbon emissions. The recycled rubber tires provided about a 20% carbon footprint advantage over coal, in energy recovery, but tires had substantially more carbon emissions than other fossil fuels. Oil pollution of marine environments is becoming a more serious issue with the growth of the off shore petroleum industry and the necessity of marine oil transportation. One of the methods to solve this problem is by using oil absorptive resins, which can collect and remove the oil spilled on water, but the high cost of the absorptive materials limits their applications. Recycled rubber from waste tyres has an intrinsic ability to absorb hydrocarbons. However, this absorption capacity of four to six times its weight is significantly lower than other absorbent materials currently in use [110]. So Wu et al. synthesized an oil absorptive material by graft copolymerization-blending method with waste tyre rubber (WTR) and 4-*tert*-butylstyrene (tBS). This can reuse the waste rubber and lowers the cost for oil absorbent production. They also investigated the quantitative influence of synthesis conditions of the oil absorptive resins on their oil absorptive properties [6].Recycled rubber from scrap tyres has been demonstrated to have potential use in oil spill cleanup by Aisien et al. because of its intrinsic ability to absorb oil. The oil absorption rate is found to be increased by the increase in the number of rubber particles and also with the environmental temperature. They compared the effect of regenerated rubber as well as fresh recycled rubber and found that the former is less efficient than the latter in potential oil clean up [110].

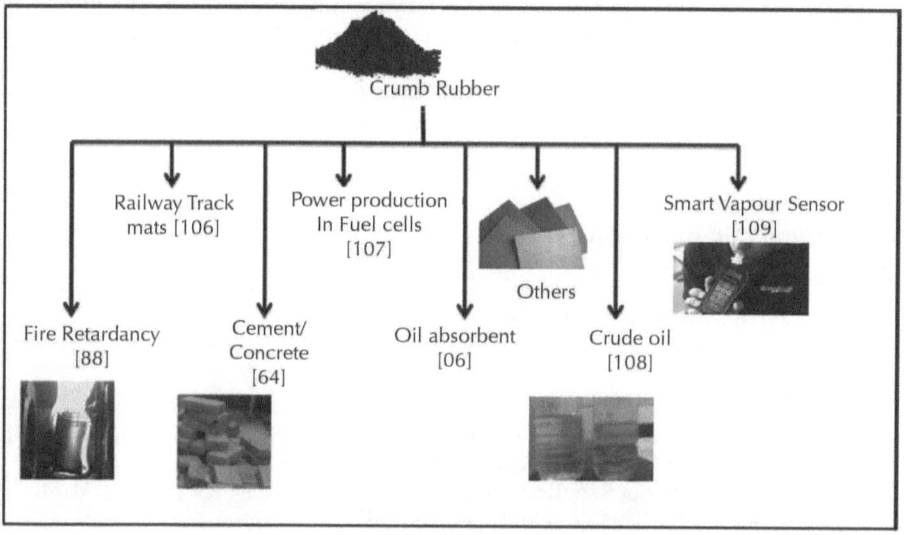

FIGURE 10 Different applications of crumb rubber.

In addition to the above methods, scrap tires have been used as a fuel for cement kiln, as feedstock for making carbon black, and as artificial reefs in marine environment [111]. But, because of the high capital investment involved in it, using tires as a fuel is technically feasible but economically not very attractive. Tripathy et al. converted scrap tires in to a semi liquid material which is proved to be effective as a filler/plasticizer for the rubber compounds. They degraded the vulcanized natural rubber and styrene butadiene rubber in nitrogen atmosphere and the resulting viscous liquid is used as a rubber plasticizer which can replace the leachable oils in rubber [112]. Scrap tire rubber is a useful component in paints and coatings for sound attenuation and also for making anti-slipping materials around pools. Tires are a source of carbon and can replace coal or coke in steel manufacturing. Tire Derived Fuel (TDF) from chipped and shredded tires is another important application. Shredded tires, also called Tire Derived Aggregate (TDA), have civil engineering applications also. It is used in backfill for walls and roadways, sub grade insulation for roads and as vibration damping material for railway lines. Apart from making reinforced asphalt and concrete, ground and crumb rubber can also be used in flooring materials, dock bumpers, railroad crossing blocks, livestock mats, rubber tiles and bricks and moveable speed bumps.

5.6 CONCLUSION

The waste management is the biggest problem that the tire industry is facing today. Tire rubber disposal methods and methods for their effective utilization are studied in detail. The physical as well as chemical ways for recycling provides better opportunity to effectively dispose the waste tire rubber. Even though the recycling is a blessing, human kind cannot escape from its adverse effects as well. Evils of recycling as well as the reusability of tire rubber are explained in detail. The tire rubber reinforced concrete, polymers and wood products are attaining great attention in today's world. Finally the various applications of waste tire products are also well explored.

KEYWORDS

- **Polymer Disposal**
- **Polymer Reuse**
- **Polymer Stabilization**
- **Rubber Recycling**
- **Rubber Reinforced Concrete**

REFERENCES

1. Sullivan, J. P. *An Assessment of Environmental Toxicity and Potential Contamination from Artificial Turf using Shredded or Crumb Rubber*. Thesis (2006).
2. Schaefer, R. and Isringhaus, R. A. Reclaimed rubber in *Rubber Technology*, (3rd Ed.). M. Morton (Ed.), Van Nostrand Reinhold, New York, pp. 505 (1987).
3. Grigoryeva, O., Fainleib, A., Starostenko, O., and Tolstov, A. Thermoplastic elastomers from rubber and recycled polyethylene: Chemical reactions at interphases for property enhancement. *Polym Int.*, **53**, 1693–1703 (2004).
4. Li, G., Stubblefield, M. A., Garrick, G., Eggers, J., Abadie, C., and Huang, B. Development of waste tire modified concrete. *Cement and Concrete Research*, **34**, 2283–2289 (2004a).
5. Karaagˇac, B., Sen, M., Deniz, V., and Gu¨ven, O. Recycling of gamma irradiated inner tubes in butyl based rubber compounds. *Nuclear Instruments and Methods in Physics Research B*, **265**, 290–293 (2007).
6. Wu, B. and Zhou, M. H. Recycling of waste tyre rubber into oil absorbent. *Waste Management*, **29**, 355–359 (2009).
7. Bruno, M. and Bu¨hrs, T. Extended producer responsibility in Brazil: The case of tyre waste. *Journal of Cleaner Production*, **17**, 608–615 (2009).
8. Milanez, B and Ton, B. Extended producer responsibility in Brazil: The case of tyre waste. *Journal of Cleaner Production*, **17**, 608–615 (2009).
9. Jawjit, W., Kroeze, C., and Rattanapan, S. Greenhouse gas emissions from rubber industry in Thailand. *Journal of Cleaner Production*, **18**, 403–411 (2010).
10. Adhikari, B., De, D., and Maiti, S. Reclamation and recycling of waste rubber. *Prog. Polym. Sci.*, **25**, 909–948 (2000).
11. Siddique, R. and Naik, T. R. Properties of concrete containing scrap-tire rubber – an overview. *Waste Management*, **24**, 563–569 (2004).
12. Warner, W. C. Method of devulcanization. *Rubb. Chem. Technol.*, **67**, 559 (1994).

13. Levin, V. Y., Kim, S. H., and Isayev, A. I. Vulcanization of Ultrasonically Devulcanized SBR Elastomers. *Rubb. Chem. Technol.*, **70**, 120–128 (1997).
14. Debapriya, D. and Debasish, D. Processing and Material Characteristics of aReclaimed Ground Rubber Tire Reinforced Styrene Butadiene Rubber. *Materials Sciences and Applications*, **2**, 486–496 (2011).
15. Norman, R. B. *Conserving the Future through the Recycling of Materials Using the Cryogenic Technique*. University of Wisconsin, pp. 393–396 (1973).
16. Eckart, R. Cryogenics advances ground rubber technology. *Modern Tire Dealer* (1980).
17. Maxwell, B. *US Patent*, **4**, 146, 508 (1979).
18. Zhang, X., Lu, C., Zheng, Q., and Liang, M. An environmental friendly approach for recycling of post-vulcanized fluoro elastomer scraps through high-shear mechanical milling. *Polym. Adv. Technol.*, **22**, 2104–2109 (2011).
19. Pelofsky, A. H. *US Patent*, **3**, 725, 314 (1973).
20. Okuda, M. and Hatano, Y. *Method of desulfurizing rubber by ultrasonic wave.* Yokohama Rubber Co. Ltd, JP, pp. 62,121,741 (1987).
21. Isayev, A. I., Yushanov, S. P., and Chen, J. Ultrasonic devulcanization of rubber vulcanizates. *I. Process model J. Appl. Polym. Sci.*, **59**, 803–813 (1996a).
22. Isayev, A. I, Yushanov, S. P, Kim, S. H., and Levin, V. Y. Ultrasonic devulcanization of waste rubbers: experimentation and modeling. *RheoActa*, **35**, 616–630 (1996b).
23. Isayev, A. I, Kim, S. H., and Levin, V. Y. Reclaimed SBR with superior mechanical properties. *Rubber Chem. Technol.*, **70**, 194–201 (1997).
24. Tukachinsky, A., Schworm, D., and Isayev, A. I. Devulcanization of Waste Tire Rubber by Powerful Ultrasound. *Rubber Chem. Technol.*, **69**, 92–103 (1996).
25. Levin, V. Y., Kim, S. H, Isayev, A. I, Massey, J., and Von Meerwall, E. Ultrasound Devulcanization of Sulfur Vulcanized SBR: Cross-link Density and Molecular Mobility. *Rubber Chem. Technol.*, **69**, 104–114 (1996).
26. Kim, D., Frank, J. Y. S., and Yen, T. F. Devulcanization of Scrap Tire through Matrix Modification and Ultrasonication. *Energy Sources*, **25**, 1099–1112 (2003).
27. Warner, W. C. Chemical methods of devulcanizing thermoset rubber. In *Plastics, Rubber and Paper Recycling, A Pragmatic Approach*. C. P. Rader, S. D. Baldwin, D. D. Cornell, G. D. Sadler, and R. F. Stockel (Eds.). ACS Symposium Series 609, Washington DC, pp. 245–253 (1995).
28. Zaharescu, T., Jipa, S., and Giurginca, M. Radiochemical processing of EPDM/NB blends. *J. Macromol. Sci. Pure and Appl. Chem. A*, **35**, 1093–1102 (1998).
29. Zaharescu, T., Melzer, V., and Vilcu, R. Thermal properties of EPDM/NR blends. *Polym. Deg. and Stab.*, (in press) (2001).
30. Kleps, T., Piaskiewicz, M., and Parasiewicz, W. The Use of Thermogravimetry in the Study of Rubber Devulcanization. *Journal of Thermal Analysis and Calorimetry*, **60**, 271–277 (2000).
31. Appleton, T. J., Colder, R. I., Kingman, S. W., and Lowndes, I. S. Microwave technology for energy-efficient processing of waste. *Applied Energy*, **81**, 85–113 (2005).
32. Sonnier, R., Leroy, E., Clerc, L., Bergeret, A., and Lopez-Cuesta, J. M. Compatibilisation of polyethylene/ground tyrerubberblends by g irradiation. *Polymer Degradation and Stability*, **91**, 2375–2379 (2006).
33. Siqin, D. and Chen, W. Radiation effects on HDPE/EVA blends. *J Appl. Polym. Sci.*, **86**, 553–558 (2002).
34. Sen, M., Uzun, C., Kantoglu, O., Erdogan, S. M., Deniz, V., and Gu¨ven, O. Effect of gamma irradiation conditions on the radiation-induced degradation of isobutylene e isoprene rubber. *NuclInstrum Methods Phys Res Sect B Beam Interact Mater Atoms*, **208**, 480–484 (2003).
35. Laresgoiti, M. F., Marco, I., Torres, A., Caballero, B., Cabrero, M. A., and Jesu´s Chomo´n, M. Chromatographic analysis of the gases obtainedintyre pyrolysis. *J. Anal. Appl. Pyrolysis*, **55**, 43–54 (2000).
36. Beck, M. R. and Klingensmith, W. Black filler for rubber from tire pyrolysis char. In: Rader C. P., Baldwin S. D., Cornell D. D., Sadler G. D., Stockel R. F. (Eds.), ACS Symposium Series 609, Washington, DC, 254–273 (1995).

37. Edward, C. O., Yanlong, S., Qin, J., Lian, S., Melizza, V., and Edward, M. E.. An Effective Coal Liquefaction Solvent Obtained fromthe Vacuum Pyrolysis of Waste Rubber Tires. *Energy and Fuels*, **10**, 573–578 (1996).
38. Mastral, A. M., Murillo, R., Callen, M. S., Perez-Surio, M. J., and Mayoral, M. C. Assessment of the Tire Role in Coal–Tire Hydrocoprocessing. *Energy and Fuels*, **11**(3) (1997).
39. Quek, A. and Balasubramanian, R. An algorithm for the kinetics of tire pyrolysis under different heatingrates. *Journal of Hazardous Materials*, **166**, 126–132 (2009).
40. Laresgoiti, M. F., Caballero, B. M., Marco, I. de, Torres, A., Cabrero, M. A., and Chomón, M. J. Characterization of the liquid productsobtained in tyre pyrolysis. *J. Anal. Appl. Pyrolysis*, **71**, 917–934 (2004).
41. Funazukuri, T., Takanashi, T., and wakao, N. Supercritical extraction of used automotive tire with water. *Journal of Chemical Engineering of Japan*, **20**(1) (1987).
42. Chen, D. T., Perman, C. A., Riechert, M. E., and Hoven, J. Depolymerization of tire and natural rubber using supercritical fluids. *Journal of Hazardous Materials*, **44**, 53–60 (1995).
43. Zhang, Q. and Tzoganakis, C. Devulcanization of recycled tire rubber using supercritical carbon dioxide. GPEC Paper Abstract #49 (2004).
44. Bhatti, I., Qureshi, K., and Zaqoot, H. A. *Super critical Decomposition of Scrap Tires Using Toluene*. Proceedings of the 7th IASME/WSEAS International conference on heat transfer, thermal engineering and environment (2009).
45. Shekhar, B. C. and Komer, V. A. De-Link Bulletin 1, De-Link recycling system, all you want to know. Kuala Lumpur: STI-K Polymers Sdn. Bhd., 1–15 (1995).
46. Kim, J. K. and Paglicawan, M. A. Effect of Devulcanizer on the Properties of Natural Rubber Vulcanizates. *Philippine Journal of Science*, **133**(2), 87–96 (2004).
47. Ishiaku, U. S., Chong, C. S., and Ismail, H. Determination of optimum De-Link R concentration in arecycled rubbercompound. *Polymer Testing*, **18**, 621–633 (1999).
48. Kim, J. K and Park, J. W. The Biological and Chemical Desulfurization of CrumbRubber for the Rubber Compounding. *Journal of Applied Polymer Science*, **72**, 1543–1549 (1999).
49. Bredberg, K., Persson, J., Christiansson, M., Stenberg, B., and Holst, O. Anaerobic desulfurization of ground rubber with the thermophilic archaeon Pyrococcus furiosus-a new method for rubber recycling. *Appl. Microbiol. Biotechnol.*, **55**, 43–48 (2001).
50. Christiansson, M., Stenberg, B., Wallenberg, L. R., and Holst, O. Reduction of surface sulphuruponmicrobialdevulcanization of rubbermaterials. *Biotechnology Letters*, **20**(7), 637–642 (1998).
51. Robert, A. R. and Margaret, F. R. Rubbercycle: A bioprocess for surface modification of waste tyre rubber. *Polymer Degrodafionond Stability*, **59**, 353–358 (1998).
52. Eldin, N. N. and Senouci, A. B. Rubber-tire particles as concrete aggregate. *J. Mater. Civ. Eng.*, **5**(4), 478–496 (1993).
53. Khatib, Z. K. and Bayomy, F. M. Rubberized Portland cement concrete. *J. Mater.Civ. Eng.*, **11**(3), 206–213 (1999).
54. Segre, N. and Joekes, I. Use of tire rubber particles as addition to cement paste, *Cem. Concr. Res.*, **30**(9), 1421–1425 (2000).
55. Fattuhi, N. I. and Clark, L. A. Cement based materials containing shredded scrap truck tyre rubber, Constr. *Build. Mater.*, **10**(4), 229–236 (1996).
56. Topcu, I. B. The properties of rubberised concretes. *Cem. Concr. Res.*, **25**(2), 304–310 (1995).
57. Topqu, I. B. Assessment of the brittleness index of rubberized concretes. *Cement and Concrete Research*, **27**(2), 177–183 (1997).
58. Lee, H. S., Lee, H., Moon, J. S., and Jung, H. W. Development of tire-added latex concrete. *ACI Mater. J.*, **95**(4) 356–364 (1998).
59. Topqu, I. B. and Avcular, N. Collision behaviors of rubberized concrete. *Cement and Concrete Research*, **27**(12), 1893–1898 (1997).
60. Li, G., Garrick, G., Eggers, J., Abadie, C., Stubblefield, M. A., and Pang, S. Waste tire fiber modified concrete. *Composites: Part B*, **35**, 305–312 (2004b).
61. Mindess, S. Material selection, proportioning, and quality control. *High Performance Concretes and Applications*, Edward Arnold, London, pp. 1–25 (1994).

62. Gu¨neyisi, E., Gesog˘lu, M., and O¨zturan, T. Properties of rubberized concretes containing silica fume. *Cement and Concrete Research*, **34**, 2309–2317 (2004).
63. Bignozzi, M. C. and Sandrolini, F. Tyre rubber waste recycling in self-compacting concrete. *Cement and Concrete Research*, **36**, 735–739 (2006).
64. Ling, T. C. Effects of compaction method and rubber content on the properties of concrete paving blocks. *Construction and Building Materials*, **28**, 164–175 (2012).
65. Olivaresa, H. F., Barluengaa, G., Bollatib, M., and Witoszek, B. Static anddynamicbehaviour of recycled tyre rubber-filled concrete. *CemConcr Res.*, **32**(10), 1587–1596 (2002).
66. Benazzouk, A., Douzane, O., Langlet, T., Mezreb, K., Roucoult, J. M., and Que´neudec, M. Physico-mechanical properties and water absorption of cement composite containing shredded rubber wastes. *Cement and Concrete Composites*, **29**,732–740 (2007).
67. Noriman, N. Z., Ismail, H., and Rashid, A. A. Characterization of styrene butadiene rubber/recycledacrylonitrile-butadiene rubber (SBR/NBRr) blends: The effectsofepoxidized natural rubber (ENR-50) as a compatibilizer. *Polymer Testing*, **29**, 200–208 (2010).
68. Takallou, H. B and Sainton, A. Advances in technology of asphalt paving materials containing used tire rubber. *Transportation Research*, Record no. 1339, 23–29 (1992).
69. Tortum, A., Celik, C., and Aydin, A. C. Determination of the optimum conditions for tire rubber in asphalt concrete. *Building and Environment*, **40**, 1492–1504 (2005).
70. Nakason, C., Worlee, A., and Salaeh, S. Effect of vulcanization systems onproperties and recyclability of dynamically cured epoxidizednaturalrubber/polypropylene blends. *Polym. Test.*, **27**, 858 (2008).
71. Bhowmick, A. K., Heslop, J., and White, J. R. Effect of stabilizers inphotodegradation of thermoplastic elastomeric rubber-polyethylene blends-a preliminary study. *Polym. Degrad. Stab.*, **74**(3), 513 (2001).
72. Dahlan, H. M., Khairul Zaman, M. D., and Ibrahim, A. The morphology andthermal properties of liquid natural rubber (LNR)-compatibilized60/40 NR/LLDPE blends. *Polym. Test.*, **21**(8), 905 (2002).
73. Pechurai, W., Nakason, C., and Sahakaro, K. Thermoplastic natural rubberbased on oil extended NR and HDPE blends: Blend compatibilizer, phase inversion composition and mechanical properties. *Polym. Test.*, **27**, 621 (2008).
74. Pichaiyut, S., Nakason, C., Kaesaman, A., and Kiatkamjornwong, S. (2008). Influences of blend compatibilizers on dynamic, mechanical, andmorphological properties of dynamically cured maleatednaturalrubber and high-density polyethylene blends. *Polym. Test.*, **27**, 566 (2008).
75. Pongdhorn, S, Sirisinha, C., Promsak, S., and Thaptong, P. Properties and recyclability of thermoplastic elastomer prepared fromnatural rubber powder (NRP) and high density polyethylene (HDPE). *Polymer Testing*, **29**, 346–351 (2010).
76. Ismail, H., Awang, M., and Hazizan, M. A. Effect of waste tire dust (WTD) size on the mechanical and morphological properties of polypropylene/waste tire dust (PP/WTD) blends. *Polymer Plastics Technology and Engineering*, **45**,463–468 (2006).
77. Awang, M., Ismail, H., and Hazizan, M. A. Polypropylene-basedblends containing waste tire dust: effects of trans-polyoctylene rubber (TOR) and dynamic vulcanization. *Polymer Testing*, **26**, 779–787 (2007).
78. Awang, M., Ismail, H., and Hazizan, M. A. Processing and properties of polypropylene-latex-modified waste tyre dust blends (PP/WTDML). *Polymer Testing*, **27**, 93–99 (2008).
79. Kaynak, C, Sipahi-Saglaa, E., and Akovali, G. A fractographic study on toughening of epoxy resin using ground tyre rubber. *Polymer*, **42**, 4393–4399 (2001).
80. Meszaros, L., Barany, T., and Czvikovszky, T. EB-promoted recycling of waste tire rubber with polyolefins. *Radiat. Phys. Chem.*, (2011). doi:10.1016/j.radphyschem.2011.11.058
81. Scaffaro, R., Tzankova, D. N., Nocilla, M. A., and La Mantia, F. P. Formulation, characterization and optimization of the processing condition of blends of recycled polyethylene and ground tyre rubber: Mechanical and rheological analysis. *Polym. Degrad. Stabil.*, **90**, 281–287 (2005).

82. Khaled, F. El-Nemr and Khalil, A. M. Gamma Irradiation of Treated Waste Rubber Powderand Its Composites with Waste Polyethylene. *Journal of Vinyl and additive technology*, **17**(1), 58–63 (2011).
83. Noriman, N. Z. and Ismail, H. Properties of Styrene Butadiene Rubber (SBR)/RecycledAcrylonitrile Butadiene Rubber (NBRr) Blends: The Effectsof Carbon Black/Silica (CB/Sil) Hybrid Filler and SilaneCoupling Agent, Si69. *Journal of Applied Polymer Science*, **124**, 19–27 (2012).
84. Ismail, H., Nordin, R., and Noor, A. M. The Effects of Recycle Rubber Powder (RRP) Content and Various Vulcanization Systems on Curing Characteristics and Mechanical Propertiesof Natural Rubber/RRP Blends. *Iranian Polymer Journal*, **12**(5), 373–380 (2003).
85. Naskar, A. K., Bhowmick, A. K., and De, S. K. (2001).Thermoplastic Elastomeric Composition Based on Ground Rubber Tire. *Polymer Engineering and science*. **41**(6) (2003).
86. Ismail, H. and Suryadiansyah, R. Thermoplastic elastomers based on polypropylene/naturalrubber and polypropylene/recycle rubber blends. *Polymer Testing*, **21**, 389–395 (2002).
87. Carli, L. N., Bianchi, O., Mauler, R. S., and Crespo, J. S. Accelerated Aging of Elastomeric Composites withVulcanized Ground Scraps. *Journal of Applied Polymer Science*, **123**, 280–285 (2012).
88. Matkó, S., Répási, I., Szabó, A., Bodzay, B., Anna, P., and Marosi, G. Fireretardancy and environmental assessment of rubberyblends of recycled polymers. *Express Polymer Letters*, **2**(2), 126–132 (2008).
89. Varga, C. S., Miskolczi, N., Bartha, L., and Palotas, L. Modification of the mechanical properties of rubbers by introducing recycled rubber into the original mixture. *Global Nest Journal*, **12**(4), 352–358 (2010).
90. Balasubramanian, M. Cure modeling and mechanical properties of counterrotating twin screw extruder devulcanized ground rubber tire/natural rubber blends. *J Polym Res.*, **16**, 133–141 (2009).
91. Sonnier, R., Leroy, E., Clerc, L., Bergeret, A., and Lopez-Cuesta, J. M. Polyethylene/ ground tyre rubber blends: Influence of particle morphology and oxidation on mechanical properties. *Polym. Test.*, **26**, 274–281 (2007).
92. Sonnier, R., Leroy, E., Clerc, L., Bergeret, A., Lopez-Cuesta, J. M., Bretelle, A. S., and Ienny, P. Compatibilizing thermoplastic/ground tyre rubber powder blends: Efficiency and limits. *Polym. Test.*, **27**, 901–907 (2008).
93. Fa´varo, S. L., Ganzerli, T. A., de CarvalhoNeto, A. G. V., da Silva, O. R. R. F., and Radovanovic, E. Chemical, morphological and mechanical analysis of sisal fiber-reinforced recycled high-density polyethylene composites. *Express Polym. Lett.*, **4**, 465–473 (2010).
94. Dintcheva, T. N., Mantia, L F.P., and Malatesta, V. Effect of different dispersing additives on the morphology and the properties of polyethylene-based nanocomposite films. *Express Polym. Lett.*, **5**, 923–935 (2011).
95. Burillo, G., Clough, R. L., Czvikovszky, T., Guven, O., Le Moel, A., Liu, W., Singh, A., Yang, J., and Zaharescu, T. Polymer recycling: Potential application of radiation technology. *Radiat. Phys. Chem.*, **64**, 41–51 (2002).
96. Abou Zeid, M. M., Rabie, S. T., Nada, A. A., Khalil, A. M., and Hilal, R. H. Effect of gamma irradiation on ethylene propylene dieneterpolymer rubber composites. *Nucl. Instrum. Methods B*, **266**, 111–116 (2008).
97. Noriman, N. Z., Ismail, H., and Rashid, A. A. The effects of different size of recycled acrylonitrile-butadiene rubber (NBRr) and blend ratio on curing characteristics, mechanical and morphological properties of styrene butadiene rubber/recycled acrylonitrile-butadiene rubber (SBR/NBRr) blends. Iran. *Polym. J.*, **18**(2), 139–148 (2009).
98. Zribi, K., Feller, J. F., Elleuch, K., Bourmaud, A., and Elleuch, B. Conductive polymer composites obtained from recycled poly (carbonate) and rubber blends for heating and sensing applications. *Polym. Adv. Technol.* doi: 10.1002/pat.776 (2006).
99. Bourmaud, A., Grohens, Y., Zribi, K., and Feller, J. F. Investigation of the Polycarbonate/ Crushed-Rubber Particle Interphase by Nanoindentation. *Journal of Applied Polymer Science*, **103**, 2687–2694 (2007).

100. Autay, R., Elleuch, K., Zribi, K., and Feller, J. F. Conductive eco-polymer composites: wearbe-haviour of recycled polycarbonate/crushed rubber microparticles. *Plastics, Rubber and Composites*, **40**(3), 139–146 (2011).
101. Fu, Z. *Properties and Designs of Rubber Materials*. Publish House of Chemistry Industry, Beijing (2003).
102. Jun, Z., Wang, X., Chang, J., and Zheng, K. Optimization of processing variables in wood–rubber composite board manufacturing technology. *Bioresource Technology*, **99**, 2384–2391 (2008).
103. Song, X. M. and Hwang, J. Y. Mechanical properties of composites made with wood fiber and and recycled tire rubber. *Forest Products Journal*, **51**(5), 45–51 (2001).
104. Ayrilmis, N., Buyuksari, U., and Avci, E. Utilization of waste tire rubber in manufacture of oriented strandboard. *Waste Management*, **29**, 2553–2557 (2009).
105. Buekenssome, A. G. Observations onthe recycling ofPlastics and rubber. *Conservation and Recyclins*, **1**, 247–271 (1997).
106. Lapck, L., Augustin, P., Pstek, A., and Bujnoch, L. Measurement of the dynamic stiffness of recycled rubber based railway track mats according to the DB-TL 918.071 standard, *Applied Acoustics*, **62**(9), 1123–1128 (2001).
107. Wang, H., Davidsona, M., Zuob, Y., and Rena, Z. Recycled tire crumb rubber anodes for sustainable power production in microbial fuel cells. *J. Power Sources*, doi:10.1016/j.jpowsour.2011.01.082 (2011).
108. Murillo, R., Aylon, E., Navarro, M. V., Calle′n, M. S., Aranda, A., and Mastral, A. M. The application of thermal processes to valorise waste tyre. *Fuel Processing Technology*, **87**,143–147 (2006).
109. Zribi, K., Elleuch, K., Feller, J. F., Bourmaud, A., and Elleuch, B. Eco-Plastics: Morphological and Mechanical Propertiesof Recycled Poly(Carbonate)-Crushed Rubber (rPC-CR) Blends. *Polymer Engineering and Science*. doi 10.1002/pen.20870 (2007).
110. Aisien, F. A., Hymore, F. K., and Ebewele, R. O. Potential application of recycled rubber in oil pollution control. *Environmental Monitoring and Assessment*, **85**, 175–190 (2003).
111. Paul, J. *Encyclopedia of Polymer Science and Engineering*, **14**, 787–802 (1985).
112. Tripathy, A. R., Williams, D. E., and Farris, R. J. Rubber Plasticizers From Degraded/Devulcanized Scrap Rubber: A Method of Recycling Waste Rubber. *Polymer Engineering and Science*, **44**(7), 1338–1350 (2004).

6 Conversion of Paper Mill Sludge into Absorbent—The Life Cycle Assessment

Marko Likon and Jouko Saarela

CONTENTS

6.1 INTRODUCTION

Usually, sorbent materials serves as cleaning agents for sanitation of the negative environmental impacts. In many cases, easily accessible and cheap materials are used for for this purposes. That materials include natural or agricultural products such rice, oat, wheat or flax straw, different hays, baggase, cottonseed hulls, corn cobs, and peat moss [1]. Industry and public services dealing with possibility of oil spill accidents are obligated to have sorbent materials with exact knowlege of their behaoviur in the case of their use. Ussualy, they are use synthetic absorbents because of their storage stability and exact sorbent specifications. But on the other hand, industry and public bodies have to be sustainable. Sustanibility of use of any material without knowledge of backgroud of its origin can be questionable, because the whole life cycle, including production, use ,and disposal must be take into account. From this point of view the life cycle assesment (LCA) can be valuable tool for comparison of diffrent sorbent materials.

Many materials have excellent sorbent characteristics, among them are wasted natural fibers or industrial by-products which are treated as a waste nowadays. Industrial by-products are generating through different industrial processes or energy production utilities as additional materials and industrial symbiosis theory defines undeliberately produced material as by-products or valuable raw materials which can be exploited in other industrial branches. According to modern industrial trends on eco symbiosis, the economic efficiency is key factor on planning modern technological processes. This includes decreasing the waste streams during production and use of produced waste as by-products or raw materials with higher added values.

Paper industry is a strategic industry in many countries but in the same time, production of the paper consumes high amounts of energy, chemicals, and wood pulp. The pulp and paper industry produce over 304 million tons of paper per year [2]. On average, the majority of waste generated from paper production and recycling is paper mill sludge (PMS), which is a by product of up to 23.4% per a unit of produced paper, the quantity depending on paper production process [3]. The countries joined in the Central European Paranormal Investigations (CEPI) organization itself produce more than 4.7 million tons of PMS per year and global production of PMS was predicted to rise over the next 50 years by between 48 and 86% above the current level. This represents an enormous environmental burden due as more than 69% of the generated PMS is landfill disposed [4]. Due to high organic content the landfilled PMS is subject to aerobic and anaerobic decay. In average the landfilling of 1 tons of PMS is releasing into environment approximately 2.69 tons of CO_2 and 0.24 ton of CH_4 [5].

As can be seen from Figure 1 PMS is chemically and physically complex material which can be usefully used in different industrial branches. The PMS consists as a mixture of short cellulosic fibers and inorganic fillers, such as, calcium carbonate and china clay, and residual chemicals dissolved in the water. The PMS characteristics vary depending on pulp and paper mill processes and it is very interesting composite material which can be used as a sustainable raw material for different applications [6-10].

FIGURE 1 SEM images shows complex structure of PMS (cellulose fibers and kaolin).

Promising research has been conducted to use PMS as an oil absorbent material and use of the PMS as sorbent material is well documented but currently the market was non receptive to such sorbent material due to cheap and efficient synthetic absorption material. The results of research studies have shown that PMS can be indirectly used as an active absorbent by converting it into activated carbon [11]. It can be used as binding material for the removal of heavy metals ions from water [12-15] removal of phenols [16] and as an absorbent for hard surfaces cleaning [17, 18]. A variety of the processes and different absorbent products have been developed for commercial purposes. One of the processes which have been developed for the production of a floor absorbent, in the form of a granular product is known as the KAOFIN process and it is described in US Patent 4343751 [19]. In US Patent 4374794 the sludge is evaporated, extruded into pellets and dried at temperatures ranging from 100 to 150°C, in order to form an oil absorbent material [20]. However, modern industry faces frequent and serious oil spills and subsequent sanitation demands high costs for sorbent materials. Offering a cheap and efficient natural material such PMS could become a welcome solution. The CAPS (Conversion of PMS into absorbent) is an eco-innovation solution in the "market uptake" phase, therefore, prior to expanding industrialization in Slovenia and Finland. The CAPS process uses the surpluses of the thermal energy which paper mills usually waste into the environment for sorbent production. In addition, CAPS uses paper mill waste as a secondary raw material and converts it into **a** high added value absorbent. The technology is relatively cheap, simple, and easily replicable particularly in markets with a developed paper industry. It is based on drying of PMS to the point where it can be efficiently mechanically and/or chemically treated to release cellulosic fibers from its inorganic matrix. The humidity of the deinking and primary PMS lies between 50 to 70%, whereas the content of cellulosic fibers is approximate 52%, the remainder is inorganic. After, drying between 70 and 80% of the solid content, PMS proceeds through special mechanical treatment (unraveling). This stage is crucial for the entire process due to the fact that in this section cellulose fibers are released from the inorganic matrix, which in turn allows material to float on the fluid surface. However, the mechanical treatment expanded the surface area but the sorbency was not linear with regards to the surface area (Table 1).

TABLE 1 Particle size, active surface area, and sorbency of used mineral oil vs mechanical treatment of PMS.

	Avg.particle size	Active surface area	Sorptivity
	mm	m^2/g	g/g
Untreated PMS	10	4.8096	1.23
PMS Grinded	4.18	3.2048	2.07
PMS Unraveled	1.67	36.0526	4.4
PMS Fluffed	0.7	2.9626	7.12

The differences in the texture of PMS after mechanical treatment are shown on Figure 2.

The unraveled PMS shows high efficiency for sorbing the mineral oils from water surfaces. As can be seen from Figure 3 the PMS sorbing the mineral oil within 45 s if the weight ratio between oil and PMS sorbent is 1:1. The PMS sorbing the mineral oil follows complex combinations of interparticle diffusion at the first stage of the process followed by pseudo-second order adsorption of oil into pores of the inorganic part of PMS (B). The interparticle diffusion of the mineral oil can be seen from Figure 4.

The chemical treatment (esterification and silanisation) is an option when higher absorption capacity and better floatability is required. Distributed (chemically prepared) PMS is dried at 130–150°C until the humidity oscillates between 1 and 10% to get final product with active surface area 36 m^2/g, sorption capacity up to 8 g oil/g PMS and capability to float on the water surface [21].

FIGURE 2 Raw PMS (left), unraveled PMS (middle), and fluffed PMS (right).

The produced natural absorbent may be used by the oil, chemical, logistic, and transport industries as well as public bodies like fire brigades, civil protection, and disaster relief institutions. These are institutions which need an environmental friendly, efficient, cheap, and at the same time sustainable product for cleaning of oil spills from water surfaces and/or for oil separators maintenance. Produced absorbent has calorific value around 3.8 MJ/kg and absorption capacity of up to 8 kg of oil per 1 kg of absorbent. Absorbent soaked by oil has calorific value up to 33.5 MJ/kg and can be used as high quality fuel in cogeneration processes. Incineration under controlled condition leads to the conversion of kaolin part of paper PMS into metakaolin substance in the form of vitrified granules. These vitrified granules can be used again as inert hydrophilic absorbent.

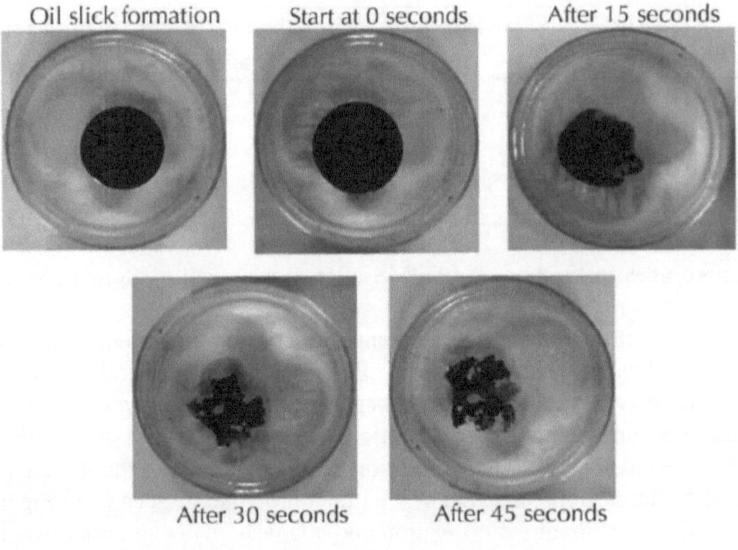

FIGURE 3 Time course of mineral oil sorbing from water surface.

In present chapter we are presenting LCA of PMS sorbent versus expanded polypropylene sorbents which is a base for discussion about sustainability of conversion of industrial waste from paper industry into high added products.

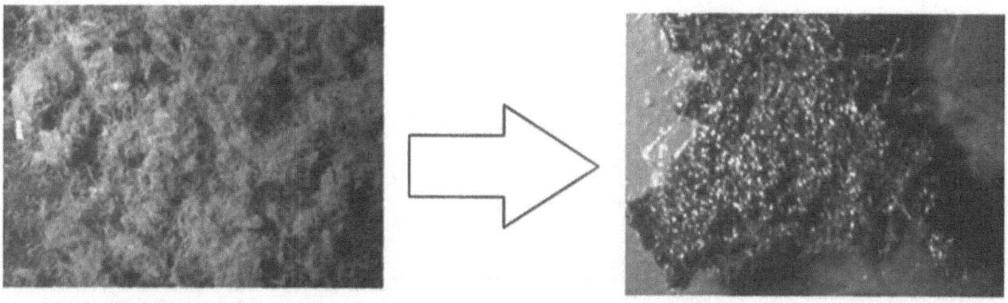

FIGURE 4 Binding of mineral oil on PMS.

6.2 METHODS

Life cycle assessments (LCA) of PMS and expanded polypropylene were accomplished by using the Eco-indicator 99(H) V2.06/Europe EI 99 H/A method with the data for polyethylene based on the BUWAL 250 database included in the software package SimaPro 7.1 and data from peer reviewed literature [22-25]. The data for the uncontrolled anaerobic decay of PMS were calculated according the [26] and the data for energy production from PMS were calculated according to [27]. The calculations of LCA for absorbent produced from PMS based on the model presenting in Figure 5.

6.2.1 LCA Calculation Boundaries

The PMS was considered as unwanted by-product in the paper and cardboard recycling processes, that is without water and energy consumption and without Greenhouse Gas (GHG) generation for its production. Expanded polypropylene (EPP) was considered as absorbent produced from non-renewable raw materials (propene). The calculations were based on the quantity of absorbent needed for absorption of 1,000 kg of oil with the following presumptions: The sorption capacity of absorbent produced from PMS and from EPP are 7 kg of oil/kg absorbent and 26 kg of oil/kg absorbent, respectively. It was considered that used absorbents were incinerated for energy recovery. Absorbed hydrophobic substances were not included in LCA calculations. The model for LCA calculation for PMS absorbent is shown on Figure 5.

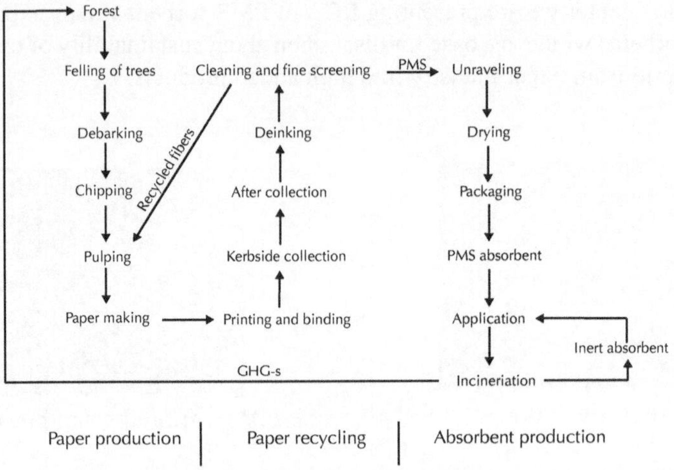

FIGURE 5 LCA calculation model for production of PMS absorbent.

6.3 DISCUSSION AND RESULTS

As can be seen from Figures 6 and 7, the production and application of PMS as sorbent for oil spill sanitation reduced carbon footprint for 2.75 times compared to the

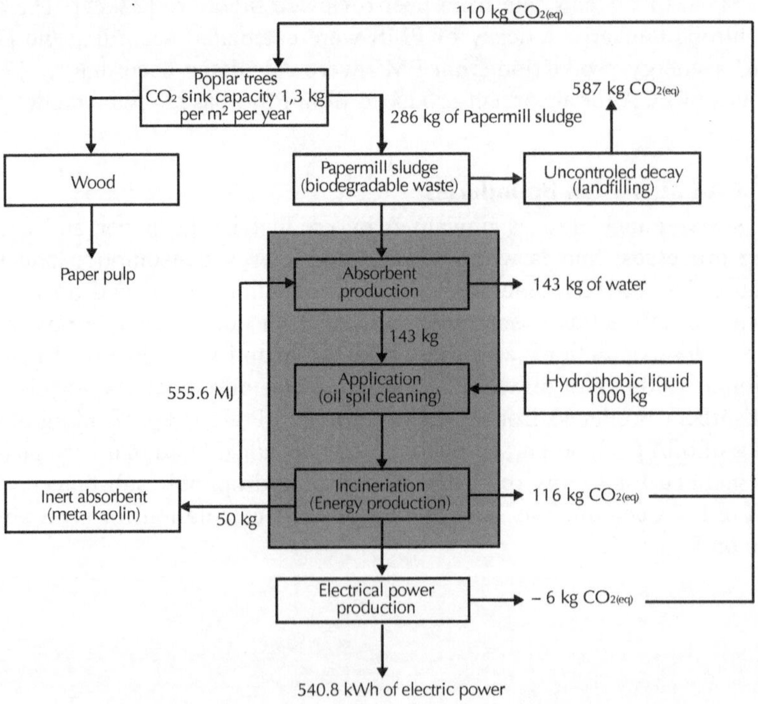

FIGURE 6 Life cycle circle for conversion of PMS into absorbent material.

production and application of EPP. In addition to this, considering the production of absorbent instead of landfilling the PMS decreased carbon footprint for additional 5.25 times (Graph 1). Altogether, production of the sorbent material from PMS instead of landfilling it and replacing the synthetic EPP absorbent with PMS absorbent for oil spill sanitations reduced carbon footprint for more than 14 times. The difference in water balance was 372.3 kg calculated on the quantity of sorbent needed for sanitation of the 1,000 kg of oil. The difference in water balance was due to consumption of the 184.3 kg of fresh water for EPP production and due to production of 143 kg of clean technological water in the process of energy recuperation during PMS sorbent production.

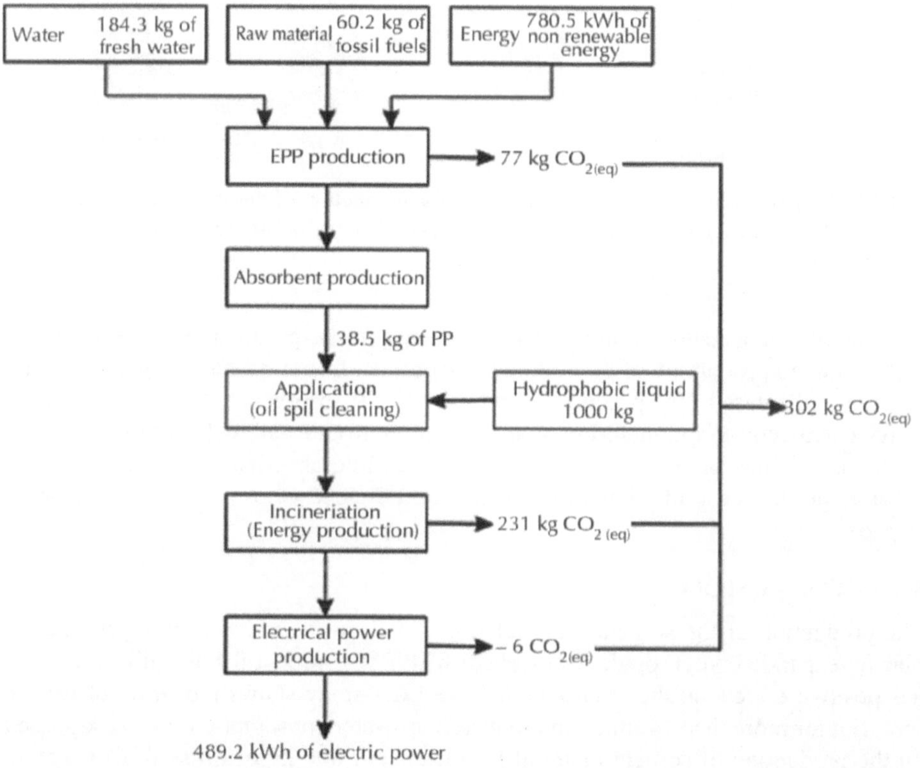

FIGURE 7 Life cycle circle production and application of EPP absorbent.

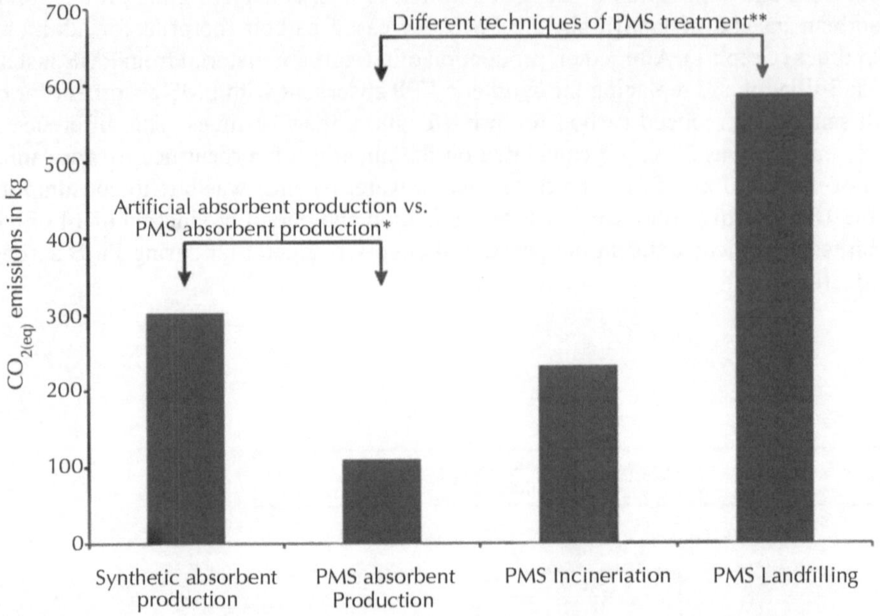

GRAPH 1 Comparison of CO2(eq) emissions for production of absorbent for sanitation of 1000 kg of oil spill (*) and comparison of CO2(eq) emissions for different tehniques for PMS management (**).

Energy balance showed the surplus of energy in the production of the PMS absorbent due to production of the combustible product from waste and negative energy balance in the case of the production and use of EPP absorbent. The later was due to heavy energy consumption during production of PP from fossil fuels. Additionally, the LCA analysis has shown that conversion of PMS into absorbent prolongs life cycle of paper products for additional two cycles and efficiently closes the life cycle circle of paper.

6.4 CONCLUSION

The production of the sorbent material from PMS, instead of landfilling it, and replacing expanded polypropylene absorbent with PMS sorbent for oil spill sanitations has positive effects on the environment. The LCA study shows reduction of carbon footprint for more than 14 times and reduction of water consumption for 372 kg based on the production of sorbent material for cleaning 1,000 kg of oil spill. Conversion of PMS into sorbent material prolongs life cycle of paper products for additional two cycles. Controlled incineration converts used sorbent into inert metakaolin product which can be further used as hydrophilic sorbent material. On that way life cycle of PMS is efficiently closed.

The modern sustainable management of production processes should be based on the industrial ecology approach, of which an essential element is the eco symbiosis theory. The LCA is valuable tool for testing the sustainability and comprehensive environmental impact of different absorbent material. The primer of production of absorbent from pulp and paper industry waste shown that the absorbent material produced from waste can be much more environmental friendly than very efficient synthetic one. With conversion of PMS into sorbent materials, the industry and public services dealing with oil, fuel or solvents spills will get cheap and sustainable sorbent material and on the other hand the pulp and paper industry will expand their product portfolio by using waste as raw material. The main drawback of placing the PMS sorbent on the market is its slow degradation in water. That obstacle can be overcome by chemical treatment, proper application, and by mixing PMS sorbent with floating materials but those processes rising the production costs and makes PMS absorbent unattractive for the market. Additional improvement must be made in the future to overcome water degradation of PMS sorbent and with success the industry can get sustainable, cheap sorbent for oil spill sanitation in unlimited quantities.

KEYWORDS

- **Central European Paranormal Investigations**
- **Chemical Treatment**
- **Life Cycle Assesment**
- **Paper Mill Sludge**
- **Sorbent Materials**

REFERENCES

1. United States Department of Agriculture. Rural Development Utilities Programs; Bulletin1724E-302; Subject: Design Guide for Oil Spill Prevention and Control at Substations. Accessed on http://www.usda.gov/rus/electric/bulletins.html (2008).
2. Lacour, P. A. Pulp and paper markets in Europe, UNECE/FAO, Url: http://www.unece.org/fileadmin/DAM/timber/docs/tc-sessions/tc-63/Presentations_MktDiscussions/Presentations_PDF/18_Lacour.pdf (accessed on Sept 26, 2005)
3. Miner, R. Environmental Considerations and Information Needs Associated With an Increased Reliance on Recycled Fiber, In Focus 95+ Proceedings. TAPPI PRESS, Atlatnta, pp. 343362 (1991).
4. Mabee,W. and Roy, D. N. Modeling the role of paper mill sludge in the organic carbon cycle of paper products. *Environmental Reviews*, **1**(11), 116 (2003).
5. Likon, M., Černec, F., Saarela, J., Zimmie, T. F., and Zule, J. Use of paper mill sludge for absorption of hydrophobic substances, 2nd International Conference on New Developments in Soil Mechanics and Geotechnical Engineering. Near East University, Nicosia, North Cyprus, pp. 526534 (2009).
6. Moo-Young, H. K. and Zimmie, T. F. Geotechnical Properties of Paper Mill Sludges for Use in Landfill Covers. *Journal of Geotechnical Engineering*, **9**(122), 768755 (1996).
7. Zule, J., Černec, F., and Likon, M. Chemical properties and biodegradability of waste paper mill sludges to be used for landfill covering. *Waste management and research*, **6**(25), 538546 (2007).
8. Černec, F., Zule, J., Može, A., and Ivanuš, A. Chemical and microbiological stability of waste sludge from paper industry intended for brick production. *Waste management and research*, **2**(23), 106112 (2005).

9. Rokainen, N., Kujala, K., and Saarela, J. Use of industrial by-products in landfill cover. Paper in Proceedings Sardinia, 12th International Waste and Landfill Symposium S. Margarita di Pula. Caligari, Italy, pp. 59 (2009).
10. Dunster, A. M. Paper sludge and paper sludge ash in Portland cement manufacture. Characterisation of Mineral Wastes, Resources and Processingtechnologies—Integrated waste management for the production ofconstruction material, Case study num. WRT 177/WR0115. Accessed on http://www.smartwaste.co.uk/filelibrary/Portland_cement_paper_sludge.pdf (2009).
11. Ben-Reuven, M. *Conversion of Paper Mill Sludge Into Pelletized, Composite Activated Sorbent; Small Business Innovation Research (SBIR) Phase I EPA*. Washington, USA. Accessed on http://cfpub.epa.gov/ncer_abstracts/index.cfm/fuseaction/display.abstractDetail/abstract/1431(1997).
12. Battaglia, A., Calace, N., and Nardi, E., Maria Petronio B.M., Pietroletti M. Paper mill sludge–soil mixture: kinetic and thermodynamic tests of cadmium and lead sorption capability. *Microchemical Journal*, **75**, 97102 (2003).
13. Calace, N., Nardi, E., Petronio, B. M., Pietroletti, M., and Tosti, G. Metal ion removal from water by sorption on paper mill sludge. *Chemosphere*, **8**(51), 797803 (2003).
14. Hea, X., Yaoa, L., Lianga, Z., and Ni, J. Paper sludge as a feasible soil amendment for the immobilization of Pb2+. *Journal of Environmental Sciences*, **22**(3), 413420 (2010).
15. Ahmaruzzaman, M. Industrial wastes as low-cost potential adsorbents for the treatment of wastewater laden with heavy metals. *Advances in Colloid and Interface Science*, **166**, 3659 (2011).
16. Calace, N., Nardi, E., Petronio, B. M., and Pietroletti, M. Adsorption of phenols by paper mill sludges. *Environmental Pollution*, **118**, 315319 (2002).
17. Lowe, H. E., Yoder, L. R., and Clayton, C. N. Nonclay oil and grease absorbent, US Patent 4734393 (1988).
18. Eifling, R. B. and Ebbers, H. J. Cellulose absorbent, US Patents 7038104 B1 (2006).
19. Naresh, K. Clay aggloomeration process. U.S. Patent 4343751 (1980).
20. Kok, J. M. Process for the preparation of a liquid-absorbing and shock-absorbing material, US Patent 4374794 (1983).
21. Likon, M., Černec, F., Svegl, F., Saarela, J., and Zimmie, T. F. Paper mill industrial waste as a sustainable source for high efficiency absorbent production. *Waste Management*, **6**(31), 13501356 (2011).
22. Felix, E., Tilley, D. R., Felton, G., and Flamino, E., Biomass production of hybrid poplar (Populus sp.) grown on deep-trenched municipal biosolids. *Ecological Engineering*, **33**, 814 (2008).
23. Binder, M. and Woods, L. *Comparative Life Cycle Assessment Ingeo™ biopolymer, PET, and PP Drinking Cups*. Final report for Starbucks Coffee Company. Five Winds and PE International. Boston USA, p. 53 (2009).
24. LyondellBasell. Polypropylene, Environmental Information Document Relevant to Australia. South Yarra Vic, Australia. Url: http://www.lyondellbasell.com/NR/rdonlyres/C2ED0A47-6430-45FA-87A4-D4018108814D/0/AusPPEnvirostatementSep09.pdf (2009).
25. Tabone, M. D., Cregg, J. J, Beckman, E. J., and Landis, A. E. Sustainability Metrics: Life Cycle Assessment and Green Design in Polymers, *Environmental Science and. Technoogy*, **44**(21), 82648269 (2010).
26. Buswell, A. M. and Mueller, H. F. Mechanism of Methane Fermentation. *Industrial Engineering Chemistry*, **44**(3), 550552 (1952).
27. Martin, F. M., Roberdo, F. G., Osado, I. I., and Ortiz, S. V. CO_2 fixation in poplar-214 plantations aimed at energy production. Council on Forest Engineering (COFE) Conference Proceedings: "*Forest Operations Among Competing Forest Uses*", Bar Harbor, US, pp.710 (2003).

7 Reuse of Natural Plant Fibers for Composite Industrial Applications

D. Saravana Bavan and G. C. Mohan Kumar

CONTENTS

7.1 INTRODUCTION

Natural fibers are considered to be the next generation fibers for the composite manufacturers, because they have the potential to replace the synthetic or the traditional fibers. In the present days composite industries are emerging from small scale to large scale sectors, which include marine, aerospace, automotive, and other military applications. Among them natural plant fibers are widely used especially in automotive and structural applications, packaging industries, and others. Various issues and aspects of natural plant fibers for industrial applications are studied. Reuse and recycle of these plant fibers for the composite technologies are focused. Developments of these natural plant fibers for the industries are highly expected in the market.

Research interest in natural fibers are growing in composite area due to the vast advantages of the fiber such as low cost, easily grown in all climatic conditions, renewable, biodegradability, low density, high strength modulus, and good thermal conductivity [1, 2]. Natural fibers are mainly classified as plant fibers, animal fibers, and mineral fibers. Plant fibers (cellulosic fibers) are further classified as bast based (e.g. hemp, flax, kenaf, jute), seed based (e.g. cotton, coir), straw based (e.g. corn, wheat, rice straw), leaf based (e.g. sisal, pine apple), and grass based (e.g. bamboo, switch grass), where as animal fibers are protein based fibers [3, 4].

The cell wall of plant fibers are mainly composed of cellulose (α-cellulose), hemicellulose, lignin, pectin, and waxes. Cellulose microfibrils are strongly surrounded and embedded by hemicelluose and lignin. Cellulose are relatively resistant to oxidizing agents [4, 5], whereas hemicelluloses are hydrophilic, easily hydrolysed in acids and they form the supportive element for cellulose. Lignins are amorphous, hydrophobic, and give stiffness to plants and pectins offer flexibility to the plant structure [4]. .These polymers differ from structure and molecular composition. Applications of natural fibers include from transportation to toy works. Major users are automotive, structural building constructions, electrical devices and appliances, packaging industries, storage devices, electronic devices, household items, and aerospace industries [2, 4, 6].

Recycling, reuse, reduce, and reform are some of major words used in the sustainable growth of industries. Whenever, a material is made, the reuse concept should be kept in mind while designing a product. In current scenario this is very useful because of the global warming effect; hence, the source of the material should be a biofriendly product and a green material. This work is focused on a reuse concept of natural fibers for industries and other applications. Some of the life cycle tool methodology, ideas, and concepts are briefed.

7.2 IMPORTANCE OF NATURAL PLANT FIBERS AND THEIR STRENGTH

Each part of the plant has its own importance, when considering the plant fiber as a composite material, and it has shown a tremendous growth in the field of manufacturing of automotive parts and building panels for engineering structural applications and the fibers behaves as an engineering material [7]. When compared to a synthetic fiber as shown in Table 1, it can be said that natural fibers are low density fibers approximately of 1.5 g/cm^3 and hence, widely used in automotive industries. Some of the disadvantages of natural fibers are low thermal properties and high moisture absorption leading to swelling of fibers. These properties can be altered by chemical treatment such as, alkalisation, acetylation, silane treatment, graft polymerization, benzoylation, and other methods [8-11].

TABLE 1 Mechanical properties of Synthetic fibers [12].

Type of Fiber	Density (g/cm^3)	Strength (MPa)	Stiffness (MPa)
Glass	2.49	2700	70
Kevlar	1.44	2800	124
Nylon 6	1.14	503690	1.82.3
Polypropylene	0.91	170325	1.62.4

TABLE 2 Mechanical properties and Chemical composition of natural fibers [2, 4, 13].

Type of Natural Fiber	Mechanical Properties			Chemical Composition		
	Density (g/cm³)	Tensile strength (MPa)	Youngs Modulus (GPa)	Cellulose (Wt %)	Hemi cellulose (Wt %)	Lignin (Wt %)
Cotton	1.51.6	287800	5.512.6	8999	36	
Jute	1.31.45	393773	1326.5	7075	1215	1015
Flax	1.50	3451100	27.6	65–85	1017	14
Sisal	1.45	468640	9.422.0	65	12	9.9
Coir	1.15	131175	46	3243	0.150.25	4045
Ramie	1.50	400938	61.412.8	68.676.2	1316	0.60.7

The plant fibers are used as reinforcement fibers for making a natural fiber composite material. The properties of plant fibers differ from plant to plant in terms of climatic conditions, place of growth, environmental conditions, type of harvest, and others [14, 15]. Hence, the mechanical, chemical properties, and other properties vary to each other as shown in Table 2.

7.2.1 Natural Fiber as a Composite Material

Composite markets are rapidly growing with newer products and with more novel machine and technology. The simple methods of processing composites are hand lay-up method followed by compression molding (sheet molding compound, bulk molding compound), pultrusion, filament winding, resin transfer molding, liquid molding, and auto clave methods [16, 17]. With advanced composite coming in to future, research should be focused on treating these natural fibers as an alternative fiber for the advanced composites. Natural fibers can be used to reinforce thermosetting and thermoplastic matrices. For high performance applications and to provide good mechanical properties thermosetting resins are used [4, 18, 19]. Based on the applications and type of work polymeric resins are selected. The below Figure 1 gives the information on each part of the plant fiber used for industrial and composite applications.

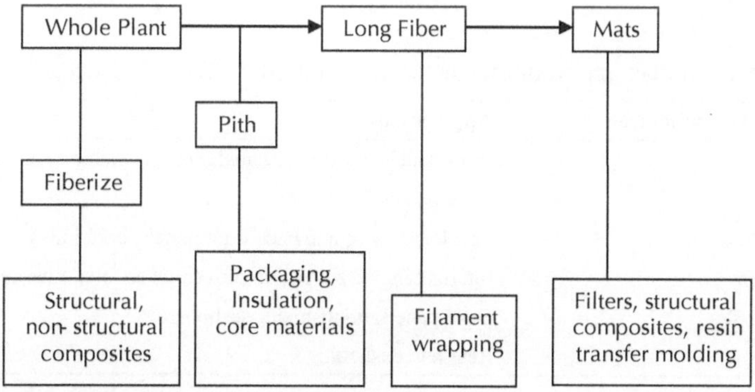

FIGURE 1 Plant fiber used for composites [20].

7.2.2 Natural Fibers in Favor of Human Kind

In Indian context, certain plant and trees are considered to be "Kalpavriksha" (also known as Kalpataru) meaning "Wish full filling divine tree", said to fulfill all desires. In Sanskrit language which means a tree that provides all the necessities of life. Coconut palm, (*Cocos Nucifera*) banyan tree, and banana plantains are some of the trees considered for *Kalpavrikshsa*. Coconut tree provides food, water, and shelter to human kind.

The coconut palm is the classic species; family from Arecaceaehad its growth in India and South Asia and later spread to European Countries. It is a fruit called *Lakshmiphal*, is used for social and religious functions in India. The tender white fruit form a part of daily diet and the coconut milk can be used for gravies and sweets. Tender coconut water is a delicious drink that contains lot of natural sugars, vitamins, and minerals. Coconut oil is used for cooking purposes in South India states like Kerala, Tamil Nadu, and Karnataka and also for hair growth. The dried coconut leaves are used for making roofing materials for the poor weaker sections society in India. The leaves are used to create products like baskets, bags, and mats. Coconut shell and husks are used for good sources of fuel and also for decorative purposes and table ware items. The mid ribs are dried, cleaned, and used to make brooms (www.coir india).

Coir fibers are thick, strong, and high abrasion resistance and water proof. They are found between the husk and the outer shell of a coconut and are resistant to damage by salt water. The fibers are narrow and hollow, with thick walls made of cellulose. The coconut tree is the national tree for Maldives country and considered to be an important plant. They are considered as a useful tree in almost all parts of the world. They are widely used for commercial, cultural, traditional, and industrial purposes.

7.2.3 Applications of Natural Fibers in Automotive and Structural Buildings

Applications of natural fiber composites in industries are dominated by automotives and building structural applications, followed by packaging industries, electronics, and others. Natural fibers uses include the non-food market product such as pulp and paper, biofuels, composite boards, textiles and so on. Door panels, head liner panel, noise insulation panels, internal engine cover, roof cover, interior insulation, windshield, and dashboard [21, 22]. Some of the main applications of the automotive sectors are listed in Table 3.

TABLE 3 Natural fiber applications in automotive industries [23].

Automotive Manufacturer	Applications
BMW	Door panels, head liner panel, noise insulation panels
Ford	Door panels, boot liner
Audi	Seat backs, side and back door panels, boot lining
Mercedes-Benz	Internal engine cover, roof cover, wheel box, interior insulation
Daimler-Chrysler	Door panels, windshield, dashboard
Renault	Rear parcel shelf

In structural building, natural fibers or the wood based fibers are mainly dedicated to the composite panel products, insulation boards, door frames, roof tiles, wall panels, roofing sheets, fibrous building panels, bricks, coir fiber, reinforced composite, and cement board [2, 24]. Agricultural residues and recycling of waste wood products are used in plyboard, particle board, medium density fiber board, and paper products [25]. Jute has high specific properties; less abrasive behavior and good dimensional stability. Jute fibers are widely used for road pavement construction [26]. Jute fabric phenol composite has been commercially used as ceiling in railway coaches replacing asbestos. Selected applications of natural fibers are shown in Figure 2.

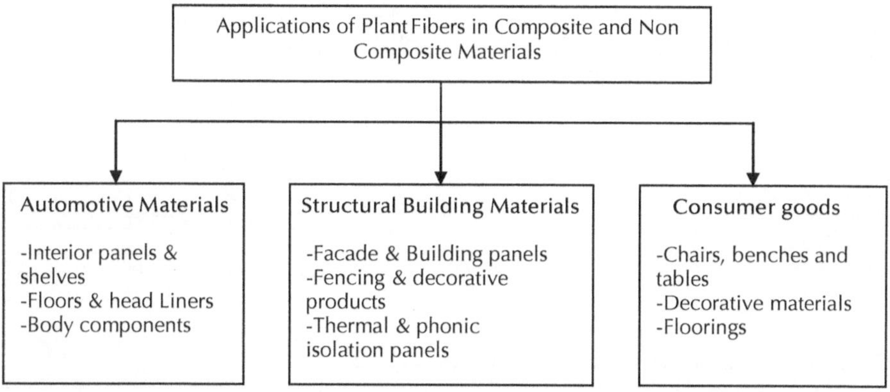

FIGURE 2 Some of the applications of plant fibers for composite materials [25].

7.2.4 Importance of Natural Plant Fibers in Each Part of the World

Coir fiber, jute fiber is widely grown in India and other South Asian Countries because of the usability, favorable climate and growing conditions. Considering coir fiber, India produces 60% of the total supply of white coir fiber whereas Sri Lanka produces 36% of the total brown fiber and 50% consumption of these fibers are from the Asian countries. Whereas hemp and flax plant fibers are vastly grown and used in European countries and regions of North America [27, 28]. Similarly each part of the land in the world has its own plant fiber and these fibers should be converted to a technical fiber and should be used for composite and other applications. As there is a slow rise of global warming, use of petro chemical products should be minimized and the use of a biodegradable fiber such as plant fiber should be maximized. Natural plant fiber reinforced materials offer many advantages such as abundant of fibers, recyclable, biodegradable, processed easily with minimum efforts, reduced pollution, and green house emission.

7.3 RECYCLE OF A COMPOSITE MATERIAL

This is the era of composite age; industries are highly focused on developing an advanced composite material that can be suited for all conditions. Industrial applications of these composite materials are increasing forever. There are different type of composite such as carbon fiber composites, glass fiber composites, hybrid fiber composites, sandwich composites, carbon nanotube reinforced composites, and natural fiber composites. Recycling of these composite materials is a tedious and lengthy task but can be achieved at the right time. Figure 3 shows the different stages of a material right from raw material extraction followed by material processing, fabrication, use, disposal and waste with recycling, land fill, and incineration methods. Figure 4 shows the waste management hierarchy, it clearly shows the different stages of order like pollution prevention, waste management, and disposal of the item. Reuse and recycle concept in the path of waste management.

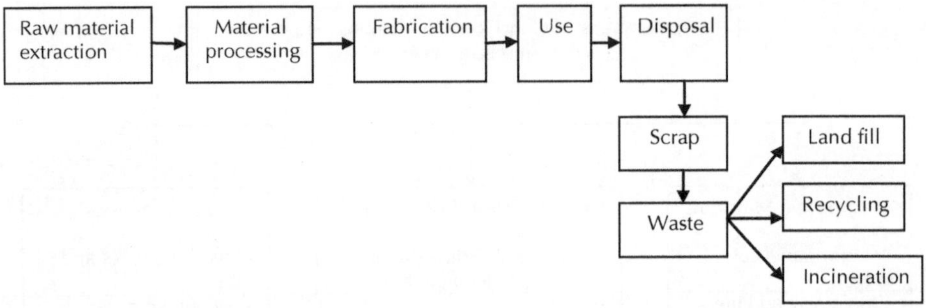

FIGURE 3 Stages of material from initial to final phase.

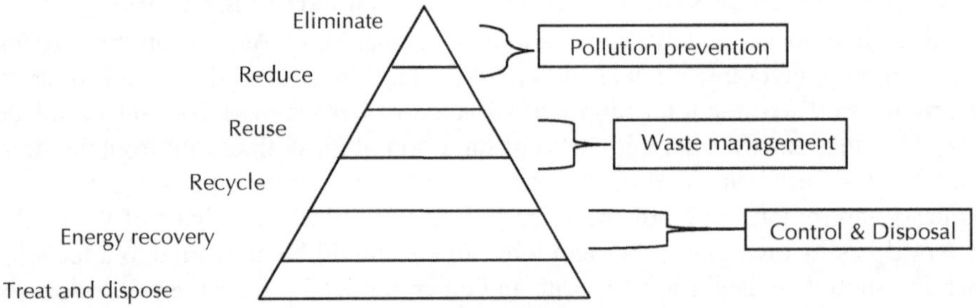

FIGURE 4 Waste management hierarchy.

Figure 5 depicts the life cycle process of a general material from initial product manufacturing to end product followed by consumer use, disposal, and recycling.

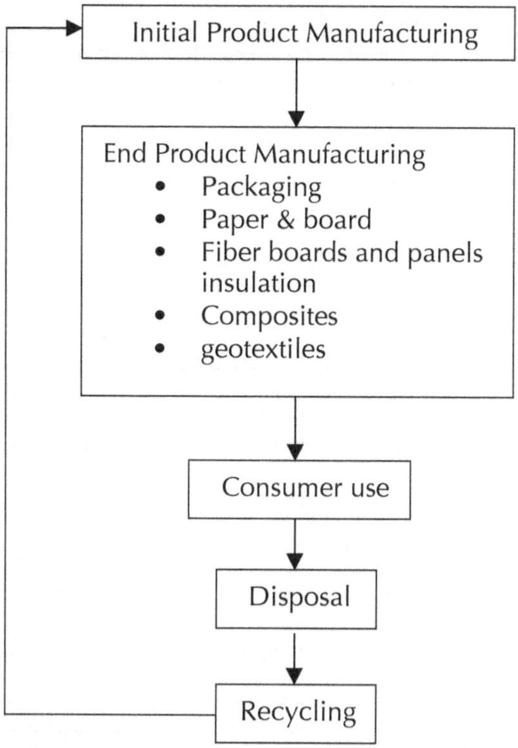

FIGURE 5 Life cycle of a general material.

In automotive industries, same pattern is followed as discussed above but the variation in recovery, reuse and remanufacture. After recovery, they are reused and sent to automobile usage, remanufacturing parts are sent to Original Equipment Manufacturer (OEM) and suppliers manufacturing, recycling products are sent to the material manufacturer. There by the usage time to delivery time is minimized. There is no dependency and each process is independent and concurrent engineering is followed. Figure 6 shows the phases of automotive cycle.

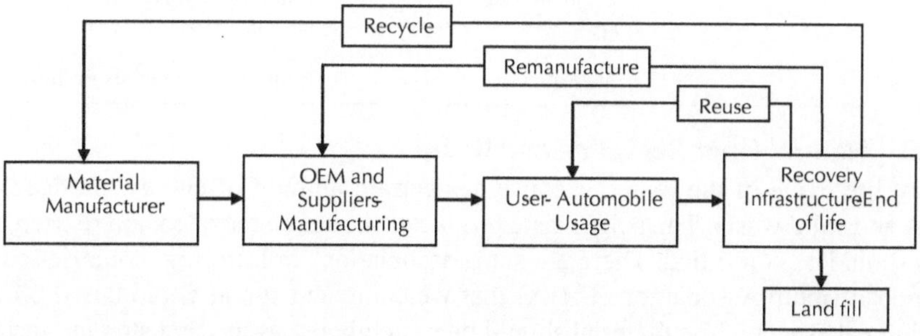

FIGURE 6 Automotive life cycle [29].

7.3.1 Products of use after Recycle

Agro waste and industrial waste can be reused after recycling. There is a tremendous potential for these waste and are used for various applications as shown in Table 4 for structural and building applications. In the similar way use of recycled composite materials for automotive sector is shown in Table 5. The applications of these recycled parts are processed in the same manner of that the original one.

TABLE 4 Types and nature of solid wastes and their recycling and utilization potentials [30].

Type of solid wastes	Source details	Recycling and utilization in structural building application
Agro wastes	Rice and wheat straw and husk, cotton stalk, ground nut shell, baggage, banana stalk, jute, sisal and other vegetable residues	Particle boards, insulation boards, wall panels, roofing sheets, fibrous building panels, bricks, coir fiber, reinforced composite, cement board
Industrial wastes	Coal combustion residues, steel slag, construction debris	Cement, bricks, blocks, concrete, wood substitute products, ceramic products

TABLE 5 Technologies for material recycling developed by Toyota Manufacturer (Courtesy: Toyota).

Type	Original Item	Recycled Part
Resin composite material	Carpet	Carpet backing
	Seat fabric	Floor silencer
	Instrument panel covering	Dash silencer
	Molded roof lining	Luggage trim
Thermosetting resin	Fiber reinforced plastic	Sunroof housing
Rubber	Weather stripping	Hose protector
Automobile shredder residue	Urethane foam and fiber	Recycled sound proofing products
	Copper wiring	Reinforcing materials for aluminium casting
	glass	Reinforcing materials for tiles

7.3.2 Natural Fiber Recycling and Reuse

Every life cycle of the material consumes certain amount of energy, produce emissions and solid waste. The solid waste produced should be recycled and re used, emission should be controlled. There are some associations in European countries such as, European Hemp Association (EIHA) that welcome and supports bio based products. Ecological aspect of the material should be remembered as the first step in fabricating a composite material.

Reuse is termed as the function of the age and durability of the material in other words extending the life of the product, where as recycle measures the material capacity to be used as are source for creation of new products. With recycling and reuse lot of energy and raw materials are saved.

Depending upon the structure of the material recycling are been carried out. In general for a polymer composite material recycling involves in mechanical recycling, chemical recycling, and Incineration (energy recovery). Mechanical recycling involves the process of shredding and granulation of waste composite material. Chemical recycling involves breaking the polymer matrix waste into hydrocarbon that is reused for other polymers and chemicals. Incineration involves in reducing the amount of waste that goes to the landfill [31, 32].

The different stages of a life cycle for a natural plant fiber composites are the agricultural stage consisting of cultivating the plants for fibers, followed by pre-processing stage which involves the preparation of fibers, decortications of fibers, cleaning the fibers, extraction and processing the fibers for the matrix. The next stage is the processing stage, which involves the preparation of natural fiber reinforced polymer composite material also termed as bio composite material. Processing of natural fiber composites remains the same procedure to that of the conventional composite material. The next stage is the consumption stage, where the users and the usage of the fibers are been done. The final level is the treatment period which involves the reuse, recycle, land fill, and incineration of the material. The life cycle stages of natural plant fiber are shown in Figure 7.

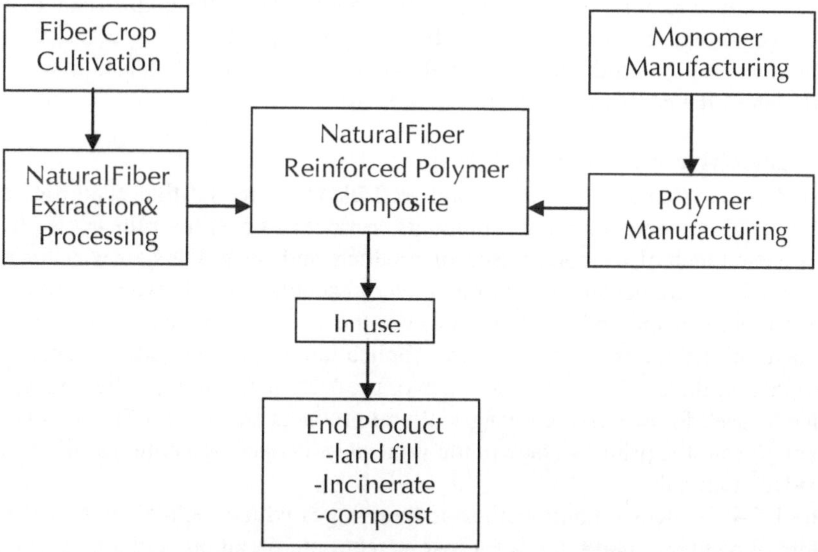

FIGURE 7 Life cycle stages of natural plant fiber composite [33].

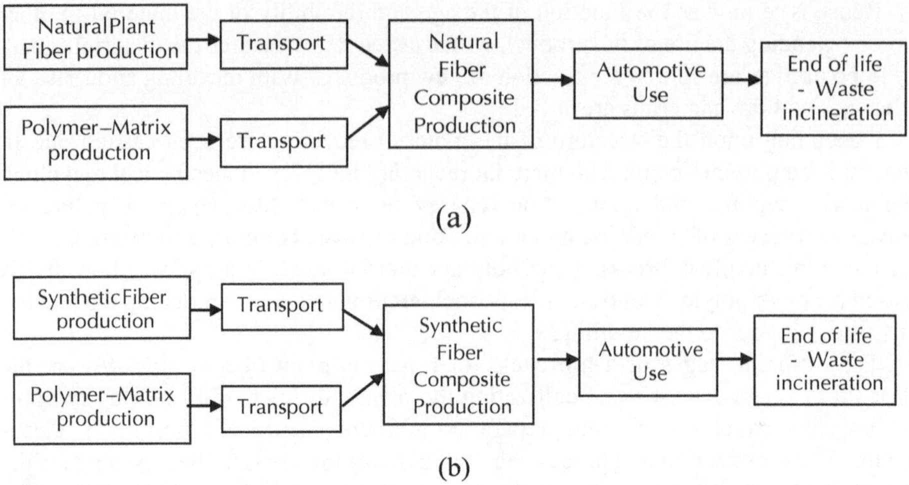

FIGURE 8 Comparison of life cycle stages of (a) natural fiber composite production to (b) synthetic fiber composite production [34].

Recycling the composites differ for thermoplastic and thermoset materials. Researcher [35] developed mechanical technique that can reclaim thermoset composites; it involves thermal process that breaks the scrap in to fibers and energy. All the constituents of the composite are reduced and brought up to recyclates, which can be obtained as mixture of polymer, fiber, and filler. Figure 8 shows the comparison between natural fiber production and synthetic fiber composite production. Processing stages are common for both productions, the difference lies only in the final stage where the degradation of the composites occur.

7.3.3 Life Cycle Assessment (LCA)

The LCA is a systematic, powerful, and well planned method that quantifies emissions and environmental impacts of products or processes over their life cycle [36-38]. They are used as tool for comparison of products and select the best way for users. The LCA of the product starts from raw material acquisition followed by production, transportation, use and end of life period. For the selection of a sustainable development material, LCA has to be followed. Application of LCA includes in quantifying and evaluating the environmental impacts of products in the product life cycle. They can also be used for evaluating the agricultural products and natural fiber composites. Evaluation from the primary stage to the final stage is rated and aimed at life cycle of the product material.

The LCA involves a combination of four stages which includes goal and scope definition, inventory analysis, impact assessment, interpretation, and improvement as shown in Figure 9. Summing briefly, it comprises of studying the boundary, analysis, and other functional items with data collection of material, energy, and emissions involved in the life cycle. Further these data are interpreted and improved for better impact assessment of the product and process [39]. The applications can be of a product development and improving the process of flow, followed by public policy making and marketing. Problems in defining the system boundary, time limitation and low spatial results were some of the back draw of LCA method.

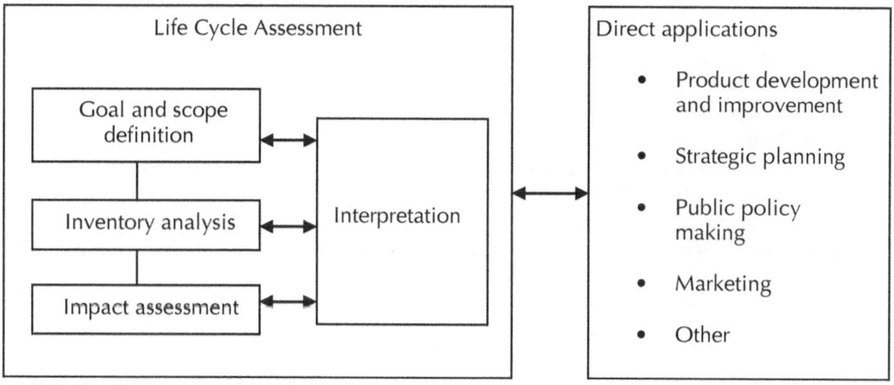

FIGURE 9 Stages of an LCA [37].

The LCA are much useful in evaluating the results of input and output statistics, and there by evaluating the environmental impacts. With the interpretation, the direct applications are interpreted to the objectives. Product life from raw material to production with end life can be dealt with LCA.

7.4 CONCLUSION

Industrial composites are emerging today at a faster rate, due to a heavy demand of composite material that are been used for various applications. But the recycling and reuse of composite materials are not paid much attention in developing and under developed countries. It is the proper and right time to rethink and reform the material that can make the environment free from oxide emissions. The world should be free from oxides of plastics; the only remedy is to develop green fiber (natural fiber) material. Minimizing of fossil fuel resources and less use of plastic resources should be borne in mind when developing a material. Recycling, Reuse, Reform and Reduce are the most common R's that should be kept in mind when developing a material. The LCA and other evaluation techniques should be created awareness to the manufacturers. It takes some more years to experience to find all the composite materials would be of natural fiber.

KEYWORDS

- **Applications**
- **Cellulose**
- **Composite**
- **Plant Fibers**
- **Reuse**

REFERENCES

1. Mohanty, A. K., Misra, M., and Hinrichsen, G. Biofibers, biodegradable polymers, and biocomposites: An overview. *Macromolecular Materials and Engineering*, **276277**, 124 (2000).
2. Mohanty, A. K., Misra, M., and Drzal, L. T. *Natural Fibers, Biopolymers, and biocomposites*. CRC Press, Taylor and Francis group, Boca Raton, FL (2005).
3. Fowler, P. A., Hughes, J. K., and Elias, R. M. Review Biocomposites: Technology, environmental credentials and market forces. *Journal of the Science of food and agriculture*, **86**(12), 17811789 (2006).
4. John, M. J. and Thomas, S. Bio fibers and Bio composites. *Carbohydrate Polymers*, **71**(3), 343364 (2008).
5. Bledzki, A. K. and Gassan, J. Composites reinforced with cellulose based fibers. *Progress in Polymer Science*, **24**(2), 221274 (1999).
6. Sen, T. and Reddy, H. N. J. Various Industrial Applications of Hemp, Kinaf, Flax and Ramie Natural Fibres. *International Journal of Innovation, Management, and Technology*, **2**(3), 192198 (2011).
7. Monteiro, S. N., Lopes, F. P. D., Barbosa, A. P., Bevitori, A. B., Da Silva, I. L. A. O. L, and Da Costa, L. L. Natural Ligno cellulosic Fibers as Engineering Materials - An Overview. *Metallurgical and Materials Transactions*, **42**(a) (2011).
8. Mohanty, A. K., Misra, M., and Drzal, L. T. Surface modifications of natural fibers and performance of the resulting biocomposites: An overview. *Composite Interfaces*, **8**(5), 31343 (2001).
9. Mwaikambo, L. Y. and Ansell, M. P. Chemical modification of hemp, sisal, jute, and kapok fibers by alkalization. *Journal of Applied Polymer Science*, **84** (12), 22222234 (2002).
10. Li, X., Tabil, L. G., and Panigrahi, S. Chemical Treatments of Natural Fiber for Use in Natural Fiber-Reinforced Composites: A Review. *Journal of Polymers and the Environment*, **15**(1), 2533 (2007).
11. John, M. J. and Anandjiwala, R. D. Recent developments in chemical modification and characterization of natural fiber-reinforced composites. *Polymer composites*, **29** (2), 187207 (2008).
12. Lilholt, H. and Lawther, J. M. Natural organic fibers, Comprehensive Composite Materials. *Vol.1: Fiber Reinforcements and General Theory of Composites*, T. W. Chou (Ed.) Elsevier, New York, pp. 303325 (2000).
13. Mwaikambo, L. Y. Review of the history, properties and application of plant fibres. *African Journal of Science and Technology Science and Engineering Series*, **7**(2) 120133 (2006).
14. Clemons, C. M. Natural Fibers, in *Functional Fillers for Plastics*. Edited by Marino Xanthos, Wiley-VCH Verlag Gmbh and Co. Weinheim. (2010).
15. Jayaraman, K. and Bhattacharyya, D. Mechanical performance of wood fiber—waste plastic composite materials, Resources. *Conservation and Recycling*, **41**, 307319 (2004).
16. Chawla, K. K. Composite materials: *Science and Engineering*, Second edition, Springer (1998).
17. Mathews, F. L and Rawlings, R. D. Composite materials: *Engineering and Science*, CRC, Wood head publishing (1999).
18. Callister, W. D. Materials Science and Engineering: An Introduction. John Wiley and Sons, New York (1999).
19. [19] Wallenberger, F. T and Weston, N. *Natural fibers, Plastics and Composites*. Kluwer academic Publishers (2003).
20. Rowell, R. M., Sanadi, A. R., Caulfield, D. F., and Jacobson, R. E. *Utilization of Natural Fibres in Plastic Composites*: Problems and Opportunities, Proceedings of International Symposium on Lignocellulosic Plastic Composites, Sao Paulo, pp. 2351 (1996).
21. Herrmann, A. S., Nickel, J., and Riedel, U. Construction materials based upon biologically renewable resources from components to finished parts. *Polymer Degradation and Stability*, **59**, 251261 (1998).
22. Holbery, J. and Houston, D. Natural-fibre-reinforced polymer composites in automotive applications. *Journal of the Minerals, Metals, and Materials society*. **58**(11), 8086 (2006). http://www.coirindia.com/

23. Suddell, B. C. and Evans, W. J. Natural Fiber Composites in Automotive Applications in Natural Fibers in Biopolymers and Their BioComposites. A. K. Mohanty, M. Misra, and L. T. Drzal (Eds.), CRC Press pp. 231259 (2005).
24. Dweib, M. A., Hu B., O' Donnell A., Shenton, H. W., and Wool, R. P. (2004). All natural composite sandwich beams for structural applications. *Composite Structures*, **63**(2), 147157.
25. Puitel, A. C., Tofanica, B. M., Gavrilescu, D., and Petrea, P. V. Environmentally Sound Vegetal Fiber Polymer Matrix Composites. *Cellulose Chemistry and Technology*, **45**(34), 265274 (2011).
26. Pandey, C. N and Sujatha, D. *Crop residues, the alternate raw materials of tomorrow for the preparation of composite board* (availed on web).
27. Sankari, H. S. Comparison of bast fibre yield and mechanical fibre properties of hemp (Cannabis sativa L.) cultivars. *Industrial Crops and Products*, **11**(1), 7384 (2000).
28. Gutierrez, A. and Rio, J. C. D. Chemical characterization of pitch deposits produced in the manufacturing of high-quality paper pulps from hemp fibers. *Bioresource Technology*, **96**, 14451450 (2005).
29. Kumar, V. and Sutherland, J. W. Sustainability of the automotive recycling infrastructure: Review of current research and identification of future challenges. *International journal of sustainable manufacturing*, **1**(1/2) (2008).
30. Pappu, A., Saxena, M., and Asolekar, S. Solid wastes generation in India and their recycling potential in building materials. *Building and Environment*, **42**(6), 23112320 (2007).
31. Morris, J. Recycling versus incineration: An energy conservation analysis. *Journal of Hazardous Materials*, **47**, 277293 (1996).
32. Allred, R. E., Doak, T. J., Coons, A. B., Newmeister, G. C., and Cochran, R. C. Tertiary recycling of cured composite aircraft parts. *Society of Manufacturing Engineers*, EM, **97**(110), X17 (1997).
33. Joshi, S. V., Drzal, L. T., Mohanty, A. K., and Arora, S. Are natural fiber composites environmentally superior to glass fiber reinforced composites? *Composites: Part A*, **35**(3), 371376 (2004).
34. Zah, R., Hischier, R., Leao, A. L., and Braun, I. Curaua fibers in the automobile industry a sustainability assessment. *Journal of Cleaner Production*, **15**(1112), 10321040 (2007).
35. Pickering, S. J. Recycling technologies for the rmoset composite materials current status. *Composites Part A Applied Science and Manufacturing*, **37**(8), 12061215 (2006).
36. Garner, A and Keoleian, G. A. *Industrial Ecology: An Introduction*. National Pollution Prevention Center for Higher Education (1995).
37. Lave, L. B. and Matthews, H. S. It is easier to say green than be green. *Technology Review*, **99**(9), 6969 (1996).
38. Vandam, J. E. G. Environmental benefits of natural fibre production and use, *Proceedings of the Symposium on Natural Fibres*, pp. 317 (2008).
39. Vandam.,J. E. G. and Bos, H. L. The Environmental Impact of Fibre Crops in Industrial Applications, Fiber crops. *www.fao.org/es/esc/common/ecg/.../environment_background.pdf.*

8 Cashew Nut Shell Liquid (CNSL): Conversion to Commercially Useful Product

Shobha Suryanarayan Sharma, Anita Nair, and K. V. Pai

CONTENTS

8.1 INTRODUCTION

The use of fossil carbon-based raw materials in the manufacture of specialty chemicals is going down because of their spiraling prices end high rate of depletion of the stocks. This necessitated a look at the renewable natural resources that can serve as alternative feedstock of raw materials for the specialty chemicals in industry. In this respect cashew nut shell liquid (CNSL) is an abundant agricultural by-product, holds considerable promise as a source of unsaturated phenol, an excellent raw material for the production of specialty chemicals. The total annual production of CNSL in India is around 15,000 tones whereas, the potential availability is around 60,000 tones and the consumption is about 11,000 tones. The rest of the CNSL is wasted or burnt. The world production is around 1.25 lakh tones [1].

The CNSL is a raw material very much utilized in green chemistry works. This oil is extracted from the cashew nut shells as a by-product in cashew industry. Natural

CNSL is a mixture of anacardic acid, cardanol, cardol, and 2-methyl-cardol in smaller quantities. The technical CNSL has been considered as one of the most promising natural source of phenols, which is used as an excellent raw material for organic and inorganic synthesis.

By socio economic concern and ISO 14001 (3R) Reduce, Recycle, Reuse the waste generated, from industries, is to be converted into valuable products. On the above concept CNSL, the by-product of cashew industry is converted into a surfactant, which finds valuable applications in pulp and paper industry as slimicide, insecticide, and penetrating aid in pulping, deinking aid, as an additive in neutral rosin sizing and as a dispersant in pigment paper coating. So, considering the great importance of this biomass and the potentiality for development of new eco-friendly compounds, this chapter narrates production of CNSL from cashew nut and its applications in brief and further synthesis of a surface active agent that is sulfonated CNSL and it is applications in paper industry.

8.1.1 History

The Cashew tree, *Anacardium occidentale L,* is a botanical species of eastern Brazil and was introduced in to other tropical countries such as India, Africa and South East Asia in the 16th Century [2]. Cashew was introduced to the Malabar Coast of India in 16th century by Spaniard and probably served as a locus of dispersal to other corners in India and South East Asia.

FIGURE 1 Cashew tree.

The Cashew tree is evergreen. It grows up to 12 m high and has spread of 25 m. It is extensive root system allows it to tolerate a wide range of moisture levels and soil types, although, commercial production is advisable only on well drained, sandy loam or red soils. Cashew tree is most frequently found in coastal areas. Cashew plant is a hardy, drought resistant plant, and comes up well in poor soil. Hence cashew is considered as a goldmine of the waste lands.

It is grown in tropical and subtropical regions of globe; however it is cultivated on a large scale in Brazil, East Africa, and southern states of India. The CNSL is traditionally obtained as a by-product during the isolation of the kernel by roasting the raw nuts.

Cashew was first introduced to India to cover the bare hills and soil conservation. It gained commercial importance in 1920, when the first shipment of cashew kernels was sent to USA. African countries produced large quantities of cashew nut but they were not processed in to consumable product due to difficulty in organizing native labor. India imports raw cashew kernels and CNSL and re-exports to the countries like USA, Canada, UK, Australia, and Russia. Though India produces only 40% of the world's production of cashew nuts, it is the major country responsible for 90% of the world production of cashew kernels. 95% or more of the apple crop is not eaten, as the taste is not popular. However, in some parts of South America and West Africa, local inhabitants regard the apple, rather than the nut kernel, as delicacy. In Brazil, the Apple is used to manufacture Jams, soft and alcoholic drinks. In Goa, in India, it is used to distill cashew liquor called feni [3].

The cashew fruit is unusual in comparison with other tree nuts since the nut is outside the fruit. The cashew apple (Figure 2) is an edible false fruit, attached to the externally born nut by a stem. The true fruit of cashew is the nut, a kidney shaped structure of approximately 23 cm in length which is attached to the end of a fleshy bulb, generally called the cashew apple. The shell comprises of some 50% of the weight of the raw nut, the kernel represents 25% and the remaining 25% consists of the natural CNSL, a viscous reddish brown liquid, within a sponge like interior. A thin skin surrounds the kernel and keeps it separated from the inside of the shell [4]. The primary products of cashew nuts are the kernels which have value as confectionery manufacture and the CNSL is traditionally obtained as a by-product during the isolation of the kernel by roasting of raw nuts.

FIGURE 2 Cashew fruit.

8.1.1.1 Processing of Cashew Nut Shell

The processing units are mainly concentrated in the states of Kerala, Tamil Nadu, Karnataka, Goa, Andhra Pradesh, Maharashtra, and Orissa.

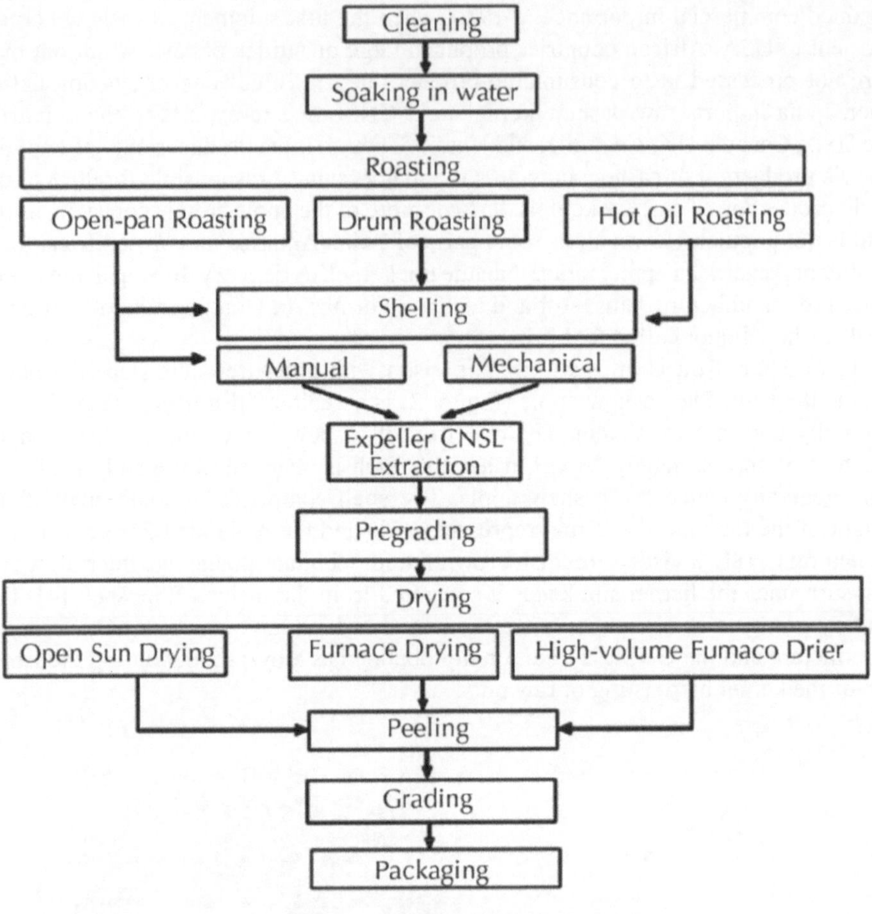

FIGURE 3 An over view of processing of cashew nut shell.

Traditionally, extraction of the kernels from the shell of the cashew nut has been a manual operation. The nut is roasted which makes the shell brittle and loosen the kernel from the inside of the shell [3]. By soaking the nuts in water, the moisture content of the kernel is raised, reducing the risk of it being scorched during roasting and making it more flexible so it is less likely to crack. The CNSL is released when nuts are roasted. It is value makes collection in sufficient quantities economically advantageous. However, for very small scale processors, this stage is unlikely to take place due to high cost of the special roasting equipment required for the CNSL collection.

Following are the steps involved in processing of cashew nut:

Cleaning: All raw nuts carry foreign matter, consisting of sand, stones, dried apple and so on. The presence of foreign matter in the roasting operation can be avoided by cleaning the nuts. The raw nuts can be sieved by hand using ¾ inch mesh sieve.

Soaking: The nuts are soaked in water to avoid scorching during the roasting operation. They are soaked for a period of not less than 4 hr in order to allow the water left on the surface of the nuts to be absorbed. The process of covering with water, draining and standing is repeated until a moisture content of 9% is reached.

Roasting: The application of heat to the nut releases the nut shell liquid and makes the shell brittle which facilitates the extraction of the kernel when breaking the shell open.

Following methods of roasting exist:

- Open pan roasting
- Drum roasting
- Hot oil roasting
- Steam roasting

Open Pan Roasting: A pan made up of mild steel, heated with nuts which are stirred constantly. The CNSL starts to exude and then ignites. After approximately 2 min, the pan is doused and charred, swollen and brittle nuts are thrown out of the pan. The moisture evaporates quickly leaving the nuts ready for shelling.

Drum Roasting: This is one of the oldest and more widely used methods. The nuts are fed into a rotating red hot drum which ignites the shell maintaining its temperature because of burning of the shell liquid. The drum is kept in rotation for 34 min and the roasted nuts are discharged from the lower end of the drum and immediately covered by ash after spraying with little water to absorb the oil on the surface. This facilitates the removal of the remaining oil on the shell. Due to draw backs of this method the method is less preferred.

Oil Bath Roasting: In this method conditioned nuts are passed through CNSL bath heated to 170–200°C by conveyor buckets for 12 min. During this period the shell gets heated rupturing the wall and releasing the oil into the bath. The oil is recovered by continuous over flow arrangement. The roasted nuts are centrifuged to remove adhering oil, cooled, and shelled by hand and leg operated shelling machines. The kernel with the adhering testa is scooped out using a sharp needle.

Steam Roasting: Steam roasting is the commonly used method by the processing units. In the case of steam roasting the raw nuts are steam cooked at about 810 kg/cm^2 pressure for about 2530 min. Then the nuts are allowed to cool for 24 hr and then taken for shelling. Shell oil can be extracted in later stages by crushing. The nuts are shelled by hand and leg operated shelling machines.

A general comparison of the above four methods would show that the oil bath method and steam roasting systems require more initial investment and higher maintenance costs. The drum roasting method is the cheapest method from the point of view of environmental pollution. The safest is the steam roasting method.

Shelling: Cashew nuts after roasting and cooling are to be shelled to remove kernels. One has to be very careful while shelling the nuts. Hands are to be protected from CNSL which is highly corrosive. Hand gloves should be used while shelling. For the same reason it is advisable to dust the nuts with wood ash. Commercial processing units use foot operated shell cutters (mechanical device) for shelling. The device consists of a pair of blade (knives) shaped in the counter of half a nut which could be operated by foot. The blades cut through the shell all around the nut leaving the kernel untouched. After shelling the kernels and shell pieces are separated manually. The nuts have to be grouped into various sizes, each size matching a pair or blades of appropriate size.

Drying: The kernels after shelling will have moisture content of more than 6%. Drying of these kernels is necessary to prevent fungus attack during subsequent storage and to facilitate peeling of testa. The kernels are to be dried to moisture content of about 45%. This is done by drying the kernels in hot chambers at 70–80°C in perforated trays of about 68 hr. Uniform drying could be achieved with a cross flow drier using forced hot air circulation through the kernel layers, in order to ensure uniform drying. The position of the trays has to be changed frequently as scorching may occur at hotter places. Excess drying of kernels makes them very brittle resulting in higher breakage. After drying, the kernels are kept in moist chambers for 24 hr which facilitates easy removal of testa and minimizes broken kernels.

Peeling: This process involves the removal to testa from the kernel. Peeling is done using a sharp knife or bamboo piece. Care has to be taken while removing the testa. If kernels are scorched it results in poor quality kernels.

Grading: Kernels are graded according to the size manually. In the international market bold whole kernels fetch premium price. The grading standards developed in India refer to white whole kernels and indicate the no. of kernels per kg of weight. The largest of kernels come in the grade W 210 (440460 /kg) and the smallest of the seven grades is W 500 (1,0001,100/kg). The white whole kernels are priced according to their size.

Packaging: As far as possible packaging material used should be eco-friendly and recyclable and containers are hermetically sealed after filling carbon dioxide.

By Products:

- Cashew apple
- Cashew apple juice
- Cashew syrup
- Cashew apple jam
- Cashew apple candy
- Cashew nut shell liquid

Cashew nut processing allows for the development of an important by product which can increase its added value. The liquid inside the shell CNSL represents 15% of gross weight and has some attractive possible medicinal and industrial uses.

Natural CNSL is manufactured by different processes.

- Hot oil process
- Kiln process
- Drum roasting process
- Expeller process
- Solvent extraction process
- Super heated steam process
- Dielectric heating radio frequency process

These methods have their own advantages and disadvantages and give CNSL with different characteristics. Hot oil method is the most common method of commercial extraction largely used in Quilon and other areas in Kerala. The nuts are passed through a bath of hot CNSL (180–200°C) itself, when the outer part of the shell bursts open and releases CNSL (50% recovery). Another 20% is extracted by passing the spent shells through an expeller or by solvent extraction technique. Expeller method is used in Mangalore and nearby areas. Some factories have introduced manually operated cutting machines in which lightly roasted nuts are cut, by keeping the kernel intact. The shells are then fed to an expeller to recover 90% of the oil. Kiln method is extensively employed Panduritti and Tamil Nadu. The nuts are shelled after sun drying or drum roasting. The CNSL obtained is however, crude and contaminated.

8.1.2 Chemistry and Composition of CNSL

Natural CNSL represents one of the major and cheapest sources of naturally occurring non-isoprenoid phenolic lipids such as, anacardic acids, cardols, cardanols, methylcardols, and polymeric materials.

Composition: Natural CNSL contains about

Anacardic acid	-	82–83.06%
Anacardanol or cardanol	-	1.6–1.77%
2 Methyl cardol	-	2.6–2.76%
Cardol	-	13.8–14.59%

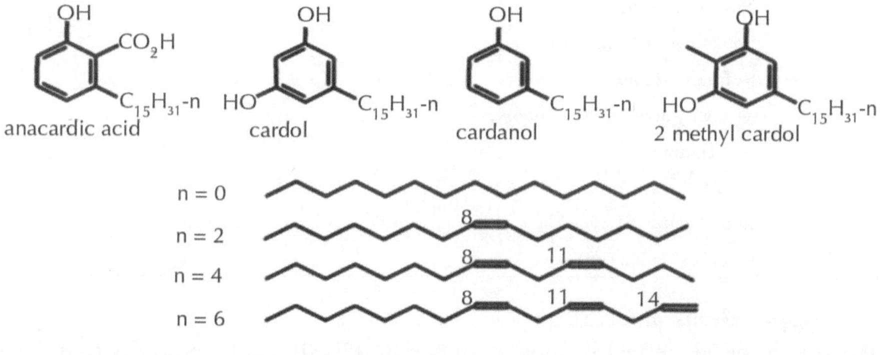

Naturally occurring non-isoprenoid phenolic lipids from *Anacardium occidentale*.

FIGURE 4 Composition of natural CNSL.

The $C_{15}H_{31-n}$ is an unsaturated side chain with one, two or sometimes three double bonds. The side chain present in anacardic acid, cardol, and cardanol is not homogeneous diolefin but a mixture of olefins of different degree of unsaturation.

Technical CNSL or cardanol (decarboxylated anacardic acid) is obtained by distilling naturally occurring phenol so called CNSL oil [5]. The first step of the process is to get the decarboxylated oil by heating the natural CNSL to a temperature of 170–175°C under reduced pressure of 3040 mm mercury.

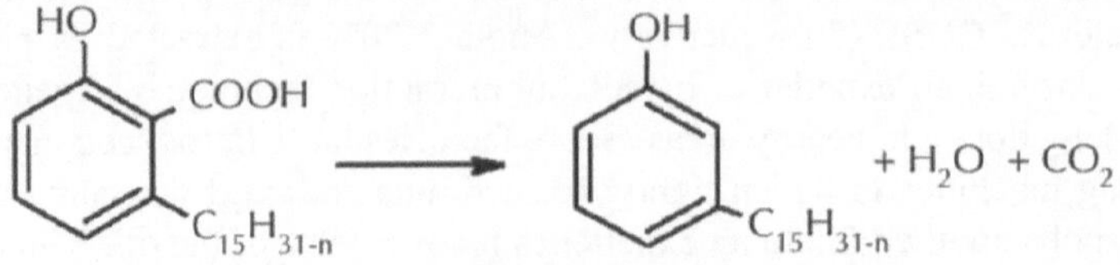

Then is subjected to fractional distillation at 200–240°C under reduced pressure not exceeding 5 mm. of mercury in the shortest possible time which gives a distillate containing cardanol, cardol, and the residual tarry matter it gives following fractions.

Cardanol	60–70%
Cardol	10–15%
Polymeric Residol	15–30%

TABLE 1 Properties of Cardanol [6].

S. No.	Particular	Grades
1	Colour	Dark brown
2	Moisture content (%)	0.0010.005
3	pH	5.68
4	Non volatile matter (%)	8385
5	Specific gravity (at 25°C)	0.9270.94
6	Refractive index	1.511.52
7	Viscosity at 300°C (Centipoises)	4565
8	Ash content	Negligible
9	Acid value (ml)	Max 5
10	Iodine value (SL I2/100 g)	210260
11	Hydroxyl value (ml)	180200

8.1.3 Applications of CNSL

In the search for cost effective modern materials, CNSL and its products have a significant role to play being renewable, it offers much advantage over synthetics. Its versatility stems from its innumerable applications in many areas. Recent research has shown that the constituents of CNSL possess special structural features for transformation into specialty chemicals and high value polymers. This involves a value addition of many orders of magnitude and the chemical transformation provides 100% chemically pure products. Thus, CNSL offers vast scope and opportunities for the production of specialty chemicals, high value products and polymers [7].

The CNSL is a unique natural source for unsaturated long chain phenols [8]. Obtained during the processing of cashew nuts, is used in the manufacture of industrially important products such as cement, specialty coatings, primers [9], paints, and varnishes, the main applications are in the polymers industry [10] based resins posses outstanding resistance to the sulfining action of mineral oils and high resistance to acids and alkalis [11].

Cardanol is a phenolic compound with a C_{15} aliphatic chain in the meta position, obtained from cashew nut shell liquid, finds many applications in form of phenol formaldehyde resins in vanishes, paints, and brakes linings [12].

Derivatives of cardanol find applications in the form of dyestuffs, plasticizers, and ion exchange resin. Chlorinated products of cardanol were found to have pesticide action.

The CNSL can be advantageously used in the manufacture of anionic and non-ionic surface active agents [8]. Like long chain fatty acids; cardanol possesses a typical lipid structure with a hydrocarbon hydrophobic group and a hydrophilic phenolic end group. This structure could be modified suitably to incorporate improved ion exchange capabilities such as introduction of a sulfonic acid group on the phenolic ring. The ion exchange resins are said to be good emulsifiers for oil-in-water and water-in-oil systems.

The CNSL forms the basic raw material for a vast number of industrially important chemicals and chemical intermediates. Patents and reports cite a number of applications such as antioxidants, bactericides, fungicides, disinfectants, insecticides, dispersing and emulsifying agents, dye stuffs [12] and so on. Hydrogenation of cardanol gives 3-pentadecylphenol which stands a good chance for industrial utilization. Reports suggest its utilization as a replacement for nonyl phenol and as a starting material for the preparation of 6-tertiarybutyl-3-pentadecylphenol and 3-pentadecylphenyl-glycedyl-ether.

It is copolymerized product with phenol and formaldehyde has been processed into specialty coatings by the Japanese. Suitable chemical modification can convert the material into plasticizers that can replace the costly petrochemical based plasticizers.

8.1.4 Synthesis of Sulfonated CNSL

Cardanol is sulfonated by using conc. H_2SO_4 at lower temperature [13]. The optimum temperature is maintained during the reaction by cooling with ice. Water can be added in the form of crushed ice, the temperature is not allowed to exceed 10°C. The product obtained is thick mass and it is kept for 30 min retention time. After 30 min the sulfonated product is washed with water to remove unreacted acid. Final product obtained after washing is a smooth shampoo like silky mass sparingly soluble in water. The sulfonated compound is then neutralized by using 10% NaOH solution to get an emulsion, buff in color with soapy smell. Sulfonation of cardanol is confirmed by I.R. spectra and the different properties are studied [6].

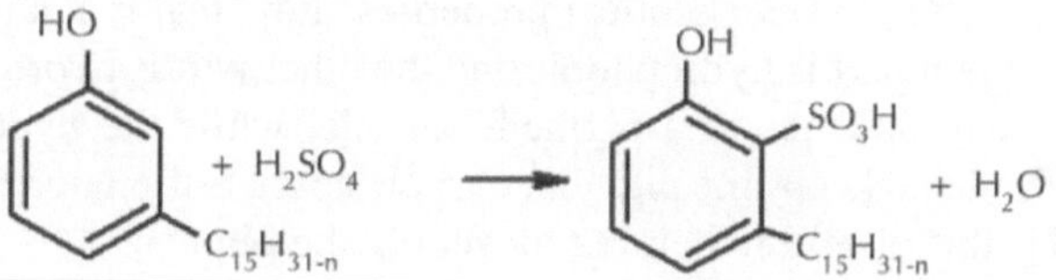

Sodium salt of sulfonated CNSL has following Physico-chemical properties [14] as shown in the table.

TABLE 2 Properties of sodium salt of sulfonated CNSL.

Properties	Values
Color	Light brown
Specific gravity (26°C)	0.9916
Moisture content (%)	25
pH	8.09.0
Surface tension (Dyne/cm)	34.94
Viscosity (cP)	7390
Iodine value (SL I_2/100 g)	180200
HLB value	68

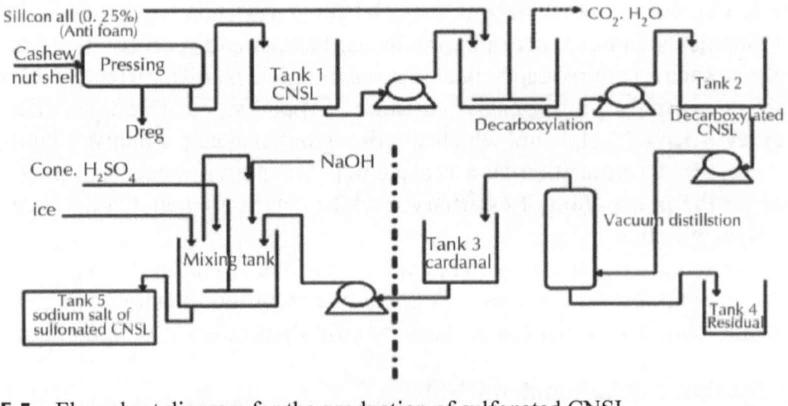

FIGURE 5 Flow sheet diagram for the production of sulfonated CNSL.

8.1.5 Applications in Pulp and Paper Industry

Dispersant in pigment coating: Sodium salt of CNSL acts as a dispersant for china clay and calcium carbonate pigments. Sodium salt of sulfonated CNCL is found to be an effective dispersant as it decreases the viscosity of the pigment slurry to the minimum value of 200 cP at the optimum dose of 0.8%. A comparative study is done with a standard dispersant (poly acrylate) commonly used in paper mills [15].

Sulfonated CNSL is a complex mixture of sulfonated cardanol and cardol. The compound has SO_3H group on its benzene ring along with OH group and a long alkyl side chain $C_{15}H_{31-n}$. The physicochemical properties show that it is a nonionic surfactant. The $C_{15}H_{31-n}$ chain part is hydrophobic and the other part is hydrophilic in nature. The hydrophobic end attaches to a pigment particle, while the hydrophilic portion sticks out in the fluid and keeps the pigment particle apart. Sulfonated CNSL is surface active agent and is thus useful in helping to wet out the pigment.

Slime control: The compound is used to study the killing efficiency on microorganisms present in white waters of hard wood based paper mill. The study included dose fixation of the compound as a slimicide, comparison with other slimicide and killing efficiencies in different white waters of paper machines. The killing efficiencies are found to be in the range of 70–90% in different paper machine white waters [16].

Natural CNSL is reported to have useful insecticidal, fungicidal, antibacterial, antidermal, and medicinal applications. But it cannot be used as slimicide because of its oily nature. Therefore, it was decided to make use of sulfonated CNSL for slime control in paper mill. The slimicide activity of the compound is mainly due to the sulfonated phenolic lipids present in it. The antimicrobial activity is retained in the molecule even after decarboxylation and sulfonation of the parent compound CNSL [17].

Penetrating aid in Kraft pulping: Sodium salt of sulfonated CNSL is used to study its application as penetrating aid, in Kraft pulping process. This study was intended to identify the potential penetrating power of sulfonated CNSL and standard compound anthraquinone both individually and in combination in Kraft pulping. The results indicated Kappa number reduction by 24 units [18].

Sulfonated CNSL increases the rate of penetration and diffusion of cooking liquor in to internal wood structure. It increases the rate of delignification, reduces cohesive forces between molecular surfaces, reduces surface tension between liquor and chips, and wets the chips surface. Thus increases penetration of the liquor inside the wood matrix. Thus mobilize and disperse the resins and fatty acids occupying wood chips flow channels. Sulfonated CNSL reduces the resin content not only in Kraft pulping but also in dissolving grade pulp. It acts as a very good chelating agent; it minimizes the metal ions of the pulp.

Additive in Rosin Sizing: The term "sizing" is a loose term. It is used to express resistance to the penetration of liquid, such as water or writing inks, Non aqueous printing inks, blood or just plain water. Papers which are resistant to liquids are called sized papers [19]. Sulfonated CNSL molecule contains phenolic, sulfonic group and a hydrophobic hydrocarbon ($-C_{15}H_{31-n}$) side chain these increases the ion exchange properties of pulp [20] and thus help in improving sizing properties. Sulfonated CNSL is a surface active agent with good dispersing properties, thus it disperses rosin properly in emulsion form [21]. These properties help in the reduction of rosin consumption in neutral rosin sizing process.

Wood preservative: The effectiveness of Sodium salt of sulfonated CNSL studied against white ants (termites both soldier and worker caste) in laboratory. It gave excellent results with mortality rate of 100%, within 10 min of exposure time, using 20% solution. The experimental design was a completely randomized design (CRD) with eight treatments, which was repeated four times. The mortality rates were recorded for different concentrations of sulfonated CNSL including controlled treatment and standard treatment.

The high mortality rates recorded at 20% concentration in very less exposure time period confirms the insecticidal property of sulfonated CNSL against termites [22]. The reactive nature and surface active properties of sulfonated CNSL constituents make it an important material for insecticide formulation. It is easy to handle in comparison with CNSL. The penetration effect is more than natural CNSL, as sulfonated

CNSL is a surface active compound. Sulfonated CNSL fulfills all the requirements of a commercial and eco-friendly wood preservative against termites.

The evaluation of results clearly shows that sulfonated CNSL can be commercially used for its surface active properties. Sulfonated CNSL is more advantageous than using plain CNSL, as it has no corrosive action to skin, and it is not sticky and oily in nature. Being surface active agent it is penetrating, solubilizing, dispersing, and wetting effects are better than CNSL.

8.2 CONCLUSION

The CNSL is the best choice for getting surface active agents. The research work related to utilization of CNSL would provide eco-friendly waste utilization process to the country as well as to the world. This chapter would help in taking laboratory results to commercial utilization in the field of pulp and paper industry, as it includes the applications in the field of pulping, sizing, slime control, insecticide,e and coating. It provides the technique for waste utilization and conversion of a by-product to commercially useful product [23]. Steps can be further taken to develop a new reliable technology which would be useful to many agro based small scale entrepreneurs to convert CNSL in to a valuable commercial product. It would increase the utilization of CNSL, part of which is either wasted or burnt.

Sulfonated cardanol and its salts can be conveniently used as detergents, emulsifying, penetrating, wetting, and solubilising agent not only in paper industry but also in other industries for example ceramic, detergent, pesticide, paint, and leather industries.

An attempt is made in this chapter to provide a technology for waste utilization and conversion of CNSL in to commercially useful product. Any work dealing with utilization of a by-product in to commercially useful product is the need of the hour and best suited to strengthen small entrepreneurship as a viable route.

KEYWORDS

- **Anacardic Acid**
- ***Anacardium Occidentale L***
- **Carbon-based Raw Materials**
- **Cardanol**
- **Cashew Nut Shell Liquid**
- **Cashew Tree**
- **Completely Randomized Design**
- **Drum Roasting**
- **Drying**
- **Expeller Method**
- **Grading**
- **Kiln Method**
- **Oil Bath Roasting**

- **Open Pan Roasting**
- **Packaging**
- **Peeling**
- **Shelling**
- **Steam Roasting**

REFERENCES

1. http://www.cardochem/com/sales@cardochem.com (accessed on October 23, 2011).
2. *Food, Chain.* The international Journal of small scale food processing: special issue dedicated to cashew processing and marketing, ITDG, may (November 28, 2001).
3. *Food, chain.* The international journal of small scale food processing. Empowering small scale cashew processors in Srilanka, ITDG, may (November 28, 1999).
4. Maria Lucilia dos Santos, Gouvan, C., and de, Magathaes. *J. Braz. Chem. Soc.*, **10**(1), 1320 (1999).
5. Tyman, J. H. P. *Chem. Ind*, **2**, 59 (1980).
6. Sharma, S. *Development of surface active agents using naturally occurring phenols from cashew nut shell liquid (CNSL)*. Report U.G.C. Minor research project under Xth plan period, India (2006).
7. Tyman, J. H. P. Synthetic and Natural Phenols. *Studies in Organic Chemistry*, **52** (1996).
8. Kumar, P. P., Paramshivappa, P. J., Vithayathil, P. J., Subra Rao, P. V., and Srinivasa Rao, A. Process for isolation of cardanol from technical cashew (Anacardium occidentale.) nut shell liquid. *J. Agric. FoodChem.*, **50**, 47054708 (2002).
9. Menon, A. R. R., Pillai, C. K. S., Sudha, J. D., and Mathew, A. G. Cashew nut shell liquid-its polymeric and other industrial products. *J. Sci. Ind. Res.*, **44**(324338), 23812383 (1985).
10. Paramshivappa, R., Phani Kumar, P., Vithayathil, P. J., and Rao, S. A. Novel method for isolation of major phenolic components from cashew (*Anacardium occidentale L*) nut shell liquid. *J. Agric. FoodChem.*, **49**, 25482551 (2001).
11. Knop, W. and Schei, *A. Chemistry and Applications of Phenolic Resin-Polymer Properties and Applications*. Springer Verlag, Berlin, Germany (1979).
12. Prabhakaran, K., Asha, N., and Pavithran, C. Cardanol as a dispersant plasticizer for analumina/toluene tape casting slip. *Journal of the European Ceramic Society*, **21**, 28732878 (2001).
13. Sharma, S., Nair, A., and Pai, K. V. Development of surface acting agents using naturally occurring phenols from CNSL by sulfonation. *IJAC journal*, **6**(2), 279286 (2010*)*.
14. Sharma, S. *Development of surface active agents using naturally occurring phenols from CNSL*. M. phil thesis Industrial Chemistry Kuvempu University, Karnataka, India (2010).
15. Sharma, S., Naik, S., Nair, A., and Pai, K. V. (In press*)*. Study of sodium salt of cashew nut shell liquid (CNSL) as an alternate dispersant in coating of paper. *IPPTA Journal*.
16. Sharma,S., Shirhatti, P. V., Nair, A., and Pai, K. V. Study of antimicrobial properties of sodium salt of sulfonated cashew nut shell liquid (CNSL) for slime control in paper mill. *IPPTA Journal*, **23**(2), 193196 (2011a).
17. Sharma, S, Nair, A., and Pai, K. V. Evaluation of antimicrobial properties of sodium salt of sulfonated cashew nut shell liquid (CNSL) as a biocide in paper mill (in preparation)
18. Sharma, S., Nair, A., and Dube, S. Development of surface acting agents using naturally occurring phenols from CNSL, a penetrating aid in Kraft pulping. *IPPTA J.* **19**(2), 125127 (2007)
19. Casey, J. P. *Pulp and Paper Chemistry and Chemical Technology*, Second edition, volume II, Paper making, Inter Science Publishers, Inc, New York, pp. 880939 (1960).
20. McLean, D. A. Ash forming constituents of insulting papers. Ind. *Eng. Chem.*, **32**(2), 209213 (1940).

21. Sharma, S., Nair, A., and Pai, K. V. Study of modified rosin sizing using sulfonated cashew nut shell liquid (CNSL) as an additive. *IPPTA Journal*, **23**(2), 109112 (2011b).
22. Sharma, S., Nair, A., and Pai, K. V. Synthesis and study of sulfonated cashew nut shell liquid (CNSL) as an effective insecticide against white ants (termites). Paper presented at UGC-SAP national symposium on Advances in Synthetic Methodologies and New Materials January 21 and 22. Paper was awarded as best paper award at Shivaji University, Kolhapur, India (2011c).
23. Sharma, S., Nair, A., and Pai, K. V. Conversion of CNSL in to commercially useful product: A technology for waste utilization. Paper presented at second International Conference on Recycling and Reuse of Materials held at Kottayam, Kerala, India (Aug 5th and 6th, 2011d).

9 Removal and Recovery of Cadmium (II) Using Immobilized Papain

Soumasree Chatterjee, Susmita Dutta, and Srabanti Basu

CONTENTS

9.1 INTRODUCTION

Cadmium is introduced to the environment through the discharge of a number of industries *viz.* electroplating, pigments, steel, batteries, solar panel and plastic and so on. Industrial effluents containing cadmium (II) need proper treatment due to the toxic effect of cadmium on living system. Bulk methods like filtration or precipitation are often practiced but they are unable to bring down the concentration to the μg/l level. The goal of this research is to remove cadmium (II) from simulated aqueous solution by immobilized papain (E.C. 3.4.22.2), a low cost proteolytic enzyme having four sulfhydryl groups which strongly bind with heavy metals. Papain has been immobilized in calcium alginate by entrapment method under the following conditions—initial concentration of sodium alginate: 20 g/l, initial concentration of calcium chloride: 20 g/l, initial concentration of papain: 25.96 g/l, pH: 7.0, temperature 35°C and hardening time: 30 min. Papain immobilized in calcium alginate under this condition has been designated as Alginate Immobilized Papain (AIP). Removal of cadmium (II) from simulated solution, kinetics and equilibrium studies has been performed using AIP. The optimum condition for removal of cadmium (II) is determined by Response Surface Methodology (RSM) using Design Expert Software 8.0.5. To get the optimized removal condition three input functions *viz.* initial concentrations of cadmium (II) weight of AIP and pH have been varied according to the experimental design as approved by software. Five gram of AIP, pH 7 and initial concentration of cadmium (II) 26.41 mg/l at 35°C have been found to be the optimum condition for removal resulting in about 93% removal of cadmium (II) from simulated solution. The adsorption kinetics of this metal has been governed by chemisorptions process as the data fit most satisfactorily to Pseudo Second Order Model (PSOM). Results reveal that the equilibrium data fit most satisfactorily with Langmuir adsorption isotherm model. Cadmium (II) is recovered from treated AIP by altering the pH of the cadmium-loaded beads. The result suggests that use of AIP could be an alternative method for removal and recovery of cadmium (II) from industrial waste.

Cadmium occupies seventh position in the list of top twenty hazardous substances as mentioned by Agency for Toxic Substances and Disease Registry (ATSDR). It is introduced to the environment through a number of industrial processes including electroplating, pigments, steel, batteries and zinc refinery, solar panel and plastic and so on. [1]. An exposure to high levels of cadmium (II) leads to an increased risk of bone fracture, cancer, kidney dysfunction, brain impairment, and hypertension [2]. According to the World Health Organization (WHO) and Indian standards, tolerance guideline for cadmium (II) in water is 0.01 mg/l. Several studies have reported higher levels of cadmium (II) in water, soil and vegetable samples. High cadmium (II) concentration ranging from 0.2 to 401.4 μg/l has been reported in the surface water in an industrial area of Tamil Nadu [3]. Another research has revealed that three locations in Alipur and two locations in the Kanjhawala of Delhi, India, were contaminated with cadmium (II) in ground water by 17 and 11 μg/l, exceeding the permissible limit in the ground water 10 μg/l set by Central Board for the Prevention and Control of Water Pollution [4]. High cadmium concentration has also been reported in the East Calcutta Wetlands, a major waste-receiving and recycling site of the city Kolkata, India [5]. It has been also found that the concentration of cadmium in the soil and vegetables of Mathura-Kanpur urban region were 6–21.53 mg/kg and 2.9 mg/kg, higher than the permissible limit (1–3 mg/kg and 1.5 mg/kg for soil and food respectively) recommended by the Indian Prevention of food Adulteration act (PFA).

It is necessary to remove cadmium (II) from industrial waste before it is discharged to the environment. Bulk methods like filtration or precipitation are often practiced but they are unable to bring down the concentration to the μg/l level. A suitable finishing step is therefore required to meet the environmental agency regulations. Presently ion exchange or chelation by crown ethers is used as the common method for removal of metals in the polishing step. The drawback of the ion exchange process lies in the non-selective nature of the process and weak binding. Crown ethers are comparatively strong binders but they often release metals slowly. This is a problem when metal recovery is required. Moreover, crown ethers are toxic in nature [6]. Bioremediation using microorganisms may be carried out but it bears the problem of disposal of filters. In this present research, an attempt has been made to use papain (E.C. 3.4.22.2), a proteolytic enzyme, for removal and recovery of cadmium (II). Being a cysteine protease papain is rich is –SH groups and can bind heavy metals including cadmium (II) [7]. This property is exploited to use papain as a tool for removal of cadmium (II) from simulated solution.

9.2 MATERIALS AND METHOD

9.2.1 Immobilization of papain

Papain was immobilized in calcium alginate beads by ionotropic gelation method following the optimum condition: initial concentration of sodium alginate: 20 g/l, initial concentration of calcium chloride: 20 g/l, initial concentration of papain: 25.96 g/l, pH: 7.0, temperature 35°C and hardening time: 30 min [8]. Papain immobilized in calcium alginate thus obtained under optimum condition was termed as AIP. Specific enzyme activity (SEA) was determined UV-Visible Spectrophotmetrically (UV 2300, TECHOM, GERMANY) extending the standard assay method for applying it on a solid immobilized sample using casein as a substrate following the procedure of [8]. The small peptides formed due to the enzymatic action of papain were measured spectrophotometrically at 280 nm.

9.2.2 Characterization of AIP

Physicochemical characterization was done in terms of temperature optima, pH optima, temperature stability and pH stability. Scanning Electron Microscopy (SEM) and

Energy dispersive X-ray spectrometry (EDS) were done to study the surface topography and presence of cadmium (II) respectively.

9.2.2.1 Determination of Temperature Optima, pH Optima, Temperature Stability and pH Stability of AIP

Temperature and pH optima of AIP were determined from its SEA at different temperatures and pH respectively using casein as the substrate following the standard procedure. Temperature and pH stability of AIP were determined from the SEA after AIP was exposed to different temperatures and pH respectively for 1 hr.

9.2.2.2 Scanning Electron Microscopy (SEM) and Energy dispersive X-ray spectrometry (EDS)

The SEM study was done to get the topological characterization of cadmium (II) adsorbed AIP and after recovery of cadmium (II) from AIP. The samples were placed on the brass stubs using double-sided adhesive tape. The SEM photographs were taken with scanning electron microscope (HITACHI–S –3000 N, JAPAN) at the required magnification at room temperature. The working distance of 25 mm was maintained and acceleration voltage used was 15 KV, with the secondary electron image (SEI) as a detector. The EDS study was done to obtain the elemental analysis of cadmium (II) adsorbed AIP to verify the binding of cadmium (II) with AIP using the above mentioned instrument.

9.2.3 Removal of Cadmium (II) Using AIP

Kinetics and equilibrium studies have been carried out to check the removal of cadmium (II) by AIP.

9.2.3.1 Kinetics Study

A series of batch experiments were carried out to evaluate the kinetics of cadmium (II) adsorption by AIP. Different initial concentration of cadmium chloride (MERCK) solution was prepared from stock solution having concentration 1,000 mg/l. Three operating parameters *viz.*, initial concentration of cadmium (II), weight of AIP and pH were varied separately in the range of 1–30 mg/l, 3–8 g, 5–9 respectively at 35°C. The volume of solution was 30 ml in each case. The solution was stirred continuously to eliminate the mass transfer resistance due to bulk diffusion. After certain interval the beads were separated from the solution and residual metal concentrations in the solutions were measured using Atomic Absorption Spectrophotometer (AAS) (JSM 6700F JEOL, Japan).

9.2.3.2 Equilibrium Study

For equilibrium study AIP was treated with 30 ml of cadmium chloride (MERCK) solution for 20 min. Initial concentration of cadmium (II) was varied in the range of 1–30 mg/l. Five gram of AIP was added to each flask and the container was placed in a BOD incubator with shaker maintained at a temperature of 35°C. The beads were separated from the solution and residual metal concentrations in the solutions were measured using AAS (JSM 6700F JEOL, Japan).

9.2.3.3 Determination of Optimum Condition for Removal of Cadmium (II) by AIP Using Response Surface Methodology

Batch mode contact device was used to study the removal of cadmium (II). The optimum condition for removal was determined by the RSM using Design Expert Software. The most popular RSM is the Central Composite Design (CCD). During optimization process three parameters *viz.* initial concentration of cadmium (II), weight of immobilized papain and the operational pH were chosen as input functions whereas percentage removal of cadmium (II) was considered as required response. The maximum (+1) and minimum (–1) level of this process variables *viz.*, initial concentration of cadmium (II), weight of *AIP* and pH were 50 and 20 mg/l, 8 and 3 g, and 10 and 4 respectively at 35°C. This resulted in 20 experiments which would have to be performed. On completion of 20 experiments and determination of residual cadmium (II) by AAS, ANOVA was employed to analyze the data for optimum removal condition. The statistical experimental design has been shown in Table 1. The spent adsorbent produced at the optimum condition was designated as Spent Alginate Immobilized Papain (SAIP) and was used for recovery of cadmium (II).

TABLE 1 Experimental design for cadmium (II) removal.

Run	Initial concentration of cadmium (II) (*A*, mg/l)	Weight of *AIP* (*B*, g)	pH (*C*)	Percentage removal (%)
1	35.00	5.50	7.00	94.3
2	50.00	3.00	4.00	45.5
3	35.00	5.50	7.00	94.3
4	35.00	1.30	7.00	46.9
5	50.00	3.00	10.00	65.7
6	20.00	3.00	10.00	78.3
7	20.00	8.00	10.00	87.4
8	35.00	9.70	7.00	97.2
9	20.00	3.00	4.00	57.5
10	20.00	8.00	4.00	69.1
11	35.00	5.50	7.00	94.3
12	35.00	5.50	1.95	38.1
13	50.00	8.00	4.00	65.6
14	35.00	5.50	7.00	94.3
15	50.00	8.00	10.00	83.8
16	9.77	5.50	7.00	87.8

TABLE 1 *(Continued)*

Run	Initial concentration of cadmium (II) (*A*, mg/l)	Weight of *AIP* (*B*, g)	pH (*C*)	Percentage removal (%)
17	35.00	5.50	12.05	62.2
18	35.00	5.50	7.00	94.3
19	35.00	5.50	7.00	94.3
20	60.23	5.50	7.00	80.5

9.2.4 Recovery of Cadmium (II)

The SAIP, obtained after filtration of cadmium (II) solution at optimum removal condition as specified by Design Expert software, was used for recovery study of cadmium (II). The SAIP was incubated in acid as well as in alkali solutions having pH 4, 7 and 9 for 30 min. The weight of SAIP and volume of solution were 5 g and 30 ml respectively in each case. After incubation, the beads were separated from the solution by filtration and concentration of cadmium (II) in the filtrate was determined using AAS.

9.3 DISCUSSION AND RESULTS

9.3.1 Immobilization of Papain

The SEA of alginate beads containing immobilized papain was found to be 785.28 (g peptide formed)/(g papain × h). The enzyme activity indicates that papain has been immobilized in alginate successfully. Removal of cadmium (II) from simulated solution, kinetics and equilibrium study has been performed using this AIP.

9.3.2 Characterization of AIP

9.3.2.1 Determination of Temperature Optima, pH optima, Temperature Stability, and pH Stability of AIP

The optimum temperature and pH for the activity of AIP are 70°C and 5.0 respectively. AIP has been found stable between a wide range of temperature (4–70°C) and pH (4–12). The experiment for removal of cadmium (II) has been carried out at 35°C to reduce the cost of heating.

9.3.2.2 Scanning Electron Microscopy (SEM) and Energy dispersive X-ray spectrometry (EDS)

The SEM images of AIP, AIP with bound cadmium (II) and AIP after recovery of cadmium (II) have been shown in Figure 1, Figure 2, and Figure 3 respectively. Comparing the topographical characterization of Figure 1 and Figure 2, it can be said that the surface of AIP bound with cadmium (II) became smoother than only AIP. It has been observed from Figure 3, that the smoothening effect of the surface was lost when cadmium (II) is recovered from AIP. Porosity on the surface of AIP reappeared after removal of cadmium. Figure 4 represents the EDS study of AIP bound with cadmium (II). The Figure shows the presence of cadmium (II) in AIP treated with cadmium (II). The result indicates that cadmium (II), when kept in contact, can bind AIP.

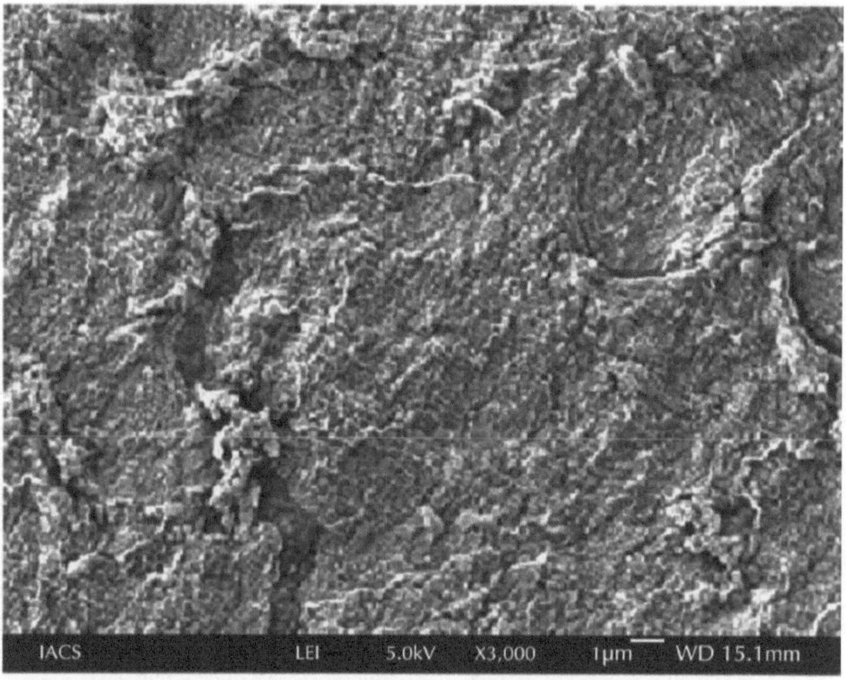

FIGURE 1 Scanning electron micrograph of Alginate Immobilized Papain (AIP).

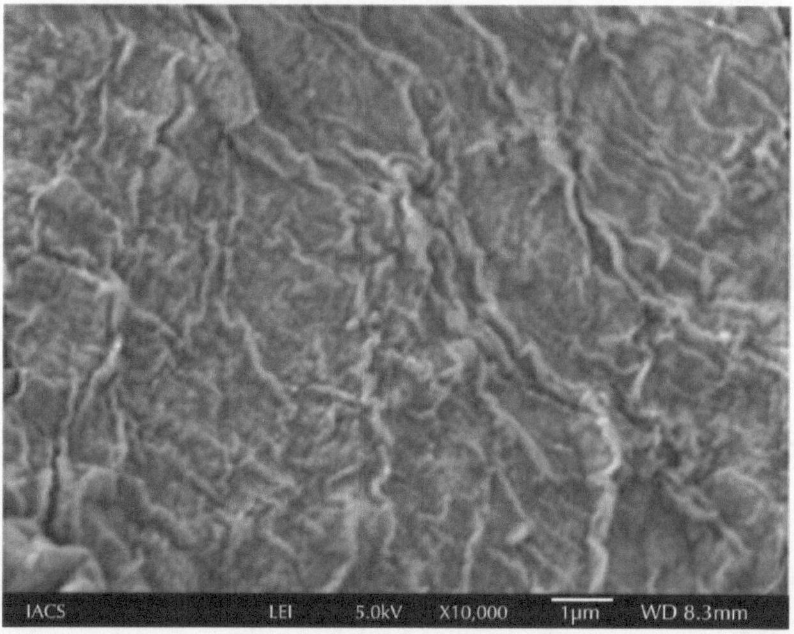

FIGURE 2 Scanning electron micrograph of AIP with bound cadmium (II).

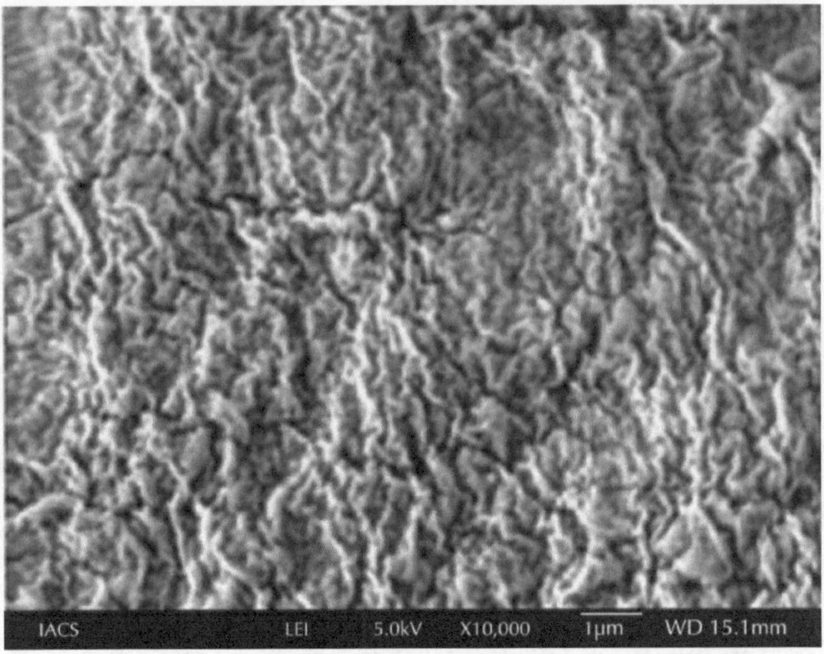

FIGURE 3 Scanning electron micrograph of AIP after recovery of cadmium (II).

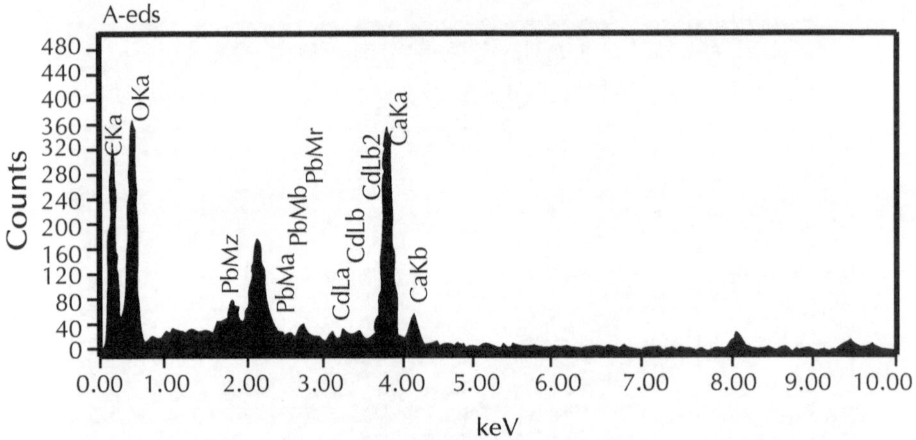

FIGURE 4 Energy dispersive X-ray spectrometry study of AIP bound with cadmium (II).

9.3.3 Removal of Cadmium (II) Using AIP

9.3.3.1 Kinetics Study

A series of batch kinetic experiments has been carried out to investigate the mechanism of adsorption process *viz.* the rate determining step and to determine the kinetic parameters. Three varying parameters *viz.*, initial concentration of cadmium (II), weight of AIP and pH have been varied in a prescribed manner during kinetic study.

Figure 5 represents the percentage removal of cadmium (II) with time considering initial concentration of cadmium (II) as varying parameter keeping weight of AIP, pH and temperature constant at 5 g, 7 and 35°C respectively. From the figure it is seen that up to 2 min of operation the rate of adsorption of cadmium (II) is very fast and is independent of the initial concentration of solution. Major part of removal is achieved within these 2 min and after then the rate of adsorption decreases and depends on residual concentration of the solution. After 10 min of operation, the percentage removal curves become parallel to abscissa indicating the attainment of equilibrium values. Maximum removal of cadmium (II) varies in the range of 84.1 and 89.3% when initial concentration changes from 1 mg/l to 30 mg/l. The modest variation in the maximum percentage removal of cadmium (II) with initial concentration of solution points out the negligible effect of this input parameter on the adsorption process. Furthermore, higher percentage removal at higher initial concentration of cadmium (II) may be attributed by the presence of greater mass transfer driving force represented as the difference between concentration of cadmium (II) in solution and that at surface. The Figure 5 also reveals that the binding sites are not saturated up to 30 mg/l of initial concentration of cadmium (II).

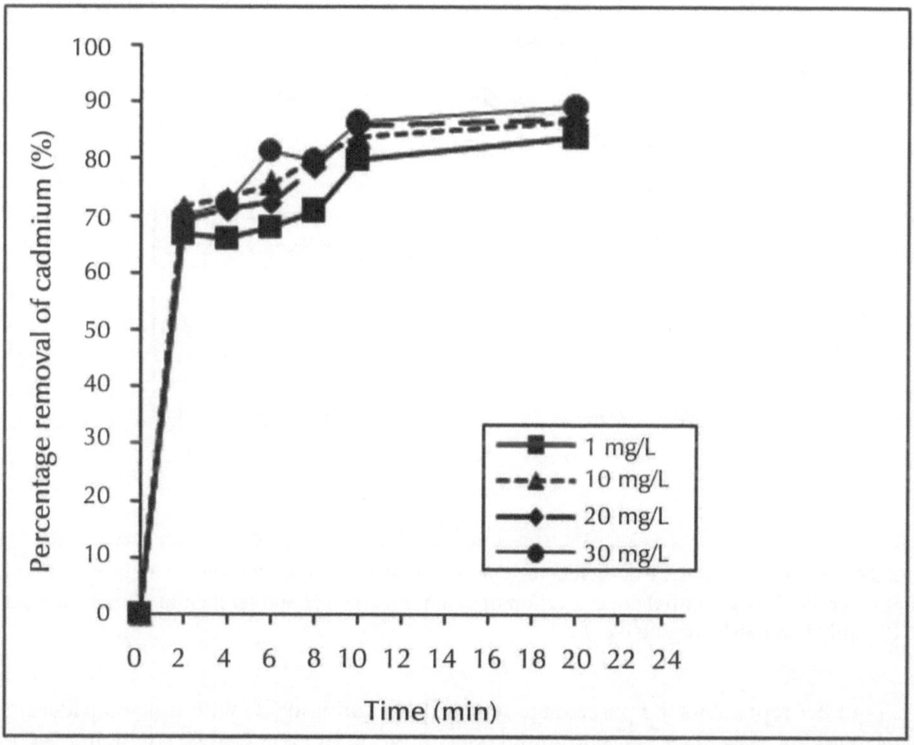

FIGURE 5 Time histories of percentage removal of cadmium (II) with initial concentration of cadmium (II) as parameter. In all cases weight of AIP, pH and temperature kept constant at 5 g, 7 and 35°C respectively.

The percentage removal of cadmium (II) with time is shown in Figure 6 considering weight of AIP as varying parameter keeping initial concentration of cadmium (II) and pH constant at 10 mg/l and 7 respectively at 35°C. In this figure also the rate of adsorption is very fast up to 2 min. Major part of removal is obtained within these 2 min and after then the rate of adsorption decreases. After 10 min of operation, the percentage removal of cadmium (II) becomes invariant with time pointing out the achievement of equilibrium. Maximum 86.9% removal is obtained when 10 mg/l solution was contacted with 5 g of AIP. Lowest removal is found in case of 3 g AIP indicating that 3 g of AIP has been saturated with cadmium (II). Under the present experimental conditions, 3 g of AIP is not adequate to remove sufficient amount of cadmium (II). Higher removal was obtained with 5 and 8 g of AIP.

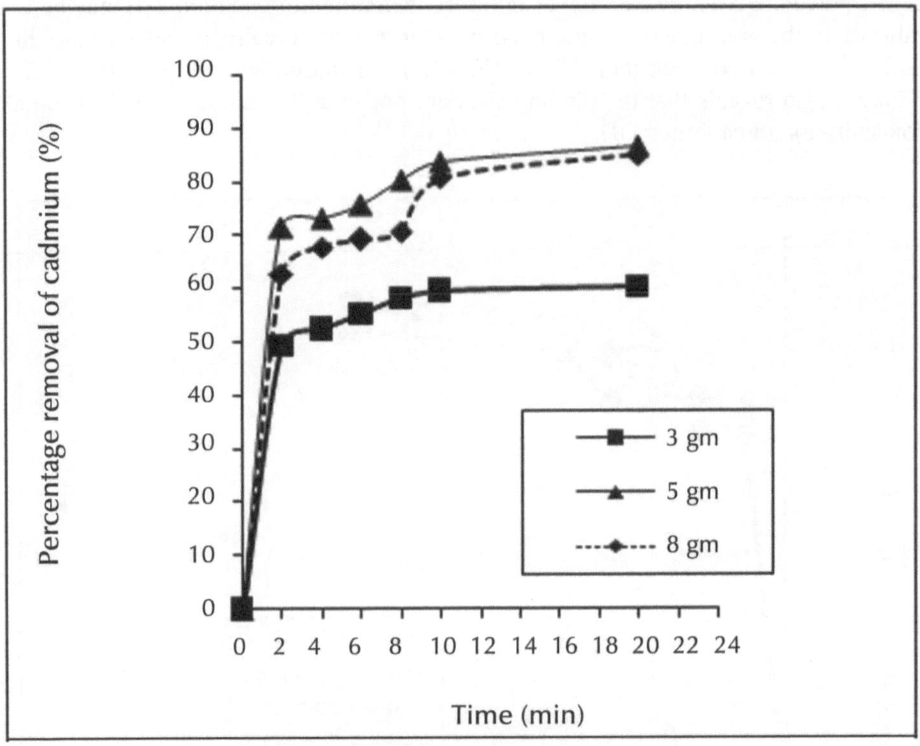

FIGURE 6 Time histories of percentage removal of cadmium (II) with weight of AIP as parameter. In all cases initial concentration of cadmium (II), pH and temperature kept constant at 10 mg/l, 7 and 35°C respectively.

Figure 7 represents the percentage removal of cadmium (II) with time considering pH as varying parameter keeping initial concentration of cadmium (II), weight of AIP and pH constant at 10 mg/l, 5 and 7 at 35°C respectively. Rate of uptake of cadmium (II) by AIP follows similar trend as observed in previous cases. From the figure, it is evident that percentage removal is strongly dependent on pH. Higher removal (86.9%)

is obtained at pH 7 whereas lower removal (44.1%) is achieved at pH 5. Papain being a sulfhydral protease gets deprotonated (~S⁻) at higher pH favorable for metal ion removal. However, slightly lower percentage removal at pH 9 (81.2%) than that at pH 7 is observed in the present case. This may be due to the conformational change of papain molecule at higher pH condition leading to a less favorable conformation for binding of cadmium. Thereby, metal removal is favored at neutral pH condition.

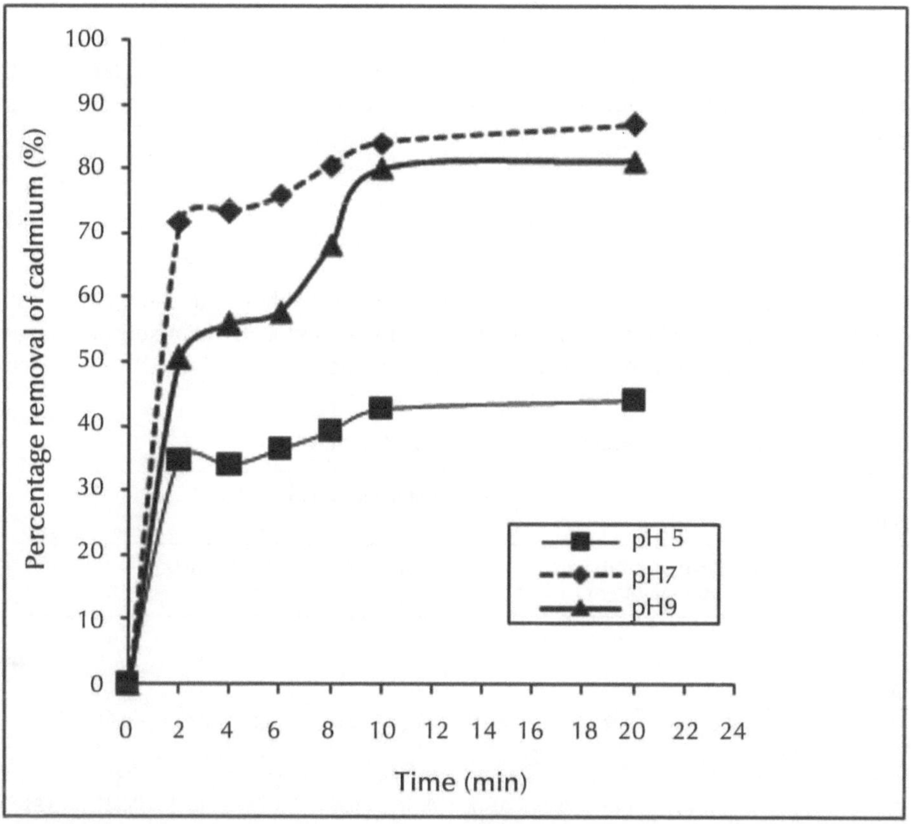

FIGURE 7 Time histories of percentage removal of cadmium (II) with pH as parameter. In all cases initial concentration of cadmium (II), weight of AIP and temperature kept constant at 10 mg/l, 5 g and 35°C respectively.

To investigate the mechanism of adsorption process three kinetic models *viz.*, Lagergren model, PSOM and Morris-Weber model have been used to test the experimental kinetic data. Lagergren first order model is based on surface reaction [9, 10]. This model can be represented as follows:

$$\log(q_e - q_t) = \log q_e - \frac{k_L' t}{2.303} \tag{1}$$

The PSOM model is chemisorptions based model. It describes that the adsorption process is of pseudo second order and may involve valence forces by sharing or through the exchange of electrons between sorbent and sorbate [9, 10]. This model can be exposed as follows:

$$\frac{t}{q_t} = \frac{1}{k'_P q_e^{\,2}} + \frac{t}{q_e} \tag{2}$$

Morris–Weber model is diffusion controlled model. It considered intraparticle diffusion as the rate limiting step when square root of time versus the uptake gives straight line passing through the origin [9, 10]. The model can be expressed by the following equation:

$$q_t = k'_M t^{\frac{1}{2}} \tag{3}$$

In all cases the experimental kinetic data fitted satisfactorily to the PSOM than any other model. It indicated that the rate of adsorption process is of chemisorption in nature. The values of kinetic parameters are shown in Table 2.

TABLE 2 Values of kinetic parameters of cadmium (II) removal.

Model equation		**Parameters varied**									
	Initial concentration of cadmium (mg/l)				**Weight of AIP (g)**			**pH**			
	1	10	20	30	3	5	8	5	7	9	
Morris Weber Model	k_M'	0.001	0.02	0.03	0.05	0.02	0.01	0.008	0.014	0.02	0.013
	R^2	0.48	0.47	0.28	0.48	0.44	0.47	0.51	0.61	0.47	0.81
Lagergren Model	k_L'	0.29	0.17	0.25	0.12	0.16	0.17	0.18	0.28	0.29	0.07
	R^2	0.49	0.29	0.31	0.19	0.34	0.29	0.27	0.16	0.17	0.11
Pseudo Second Order Model	k_P'	303.7	39.7	34.2	14.1	64.5	4.57	26.03	29.56	39.7	12.1
	R^2	0.99	0.99	0.99	0.99	0.99	0.99	0.99	0.99	0.99	0.97

9.3.3.2 Equilibrium Study

Data obtained during equilibrium study has been fitted to different adsorption isotherm models *viz.*, Langmuir model, Freundlich model and so on. The values of equilibrium parameters are calculated and shown in Table 3. Results reveal that the equilibrium data fit most satisfactorily with Langmuir adsorption isotherm model. The separation factor (R_L) expressed as $\left(\frac{1}{1+K'_L C_0}\right)$ is an important parameter of Langmuir isotherm model. The variation of R_L with initial concentration of cadmium (II) has been plotted in Figure 8. The values of R_L ranging from 0.93 to 0.58 (0 < RL < 1) with change in initial concentration cadmium (II) from 10 mg/l to 100 mg/l, implies favorable adsorption of cadmium (II) on AIP [11].

TABLE 3 Result analysis of different adsorption isotherm of cadmium (II) removal.

Adsorption isotherm model	Temperature 35°C		
	Expression	Constants	R^2
Langmuir model	$q_e = \frac{K'_L q^0 C_e}{1+K'_L C_e}$	q^0 =0.299 K'_L =0.00722	0.993
Freundlich model	$q_e = K'_F C_e^{1/n'_F}$	K'_F =0.199 n'_F=1.0073	0.911

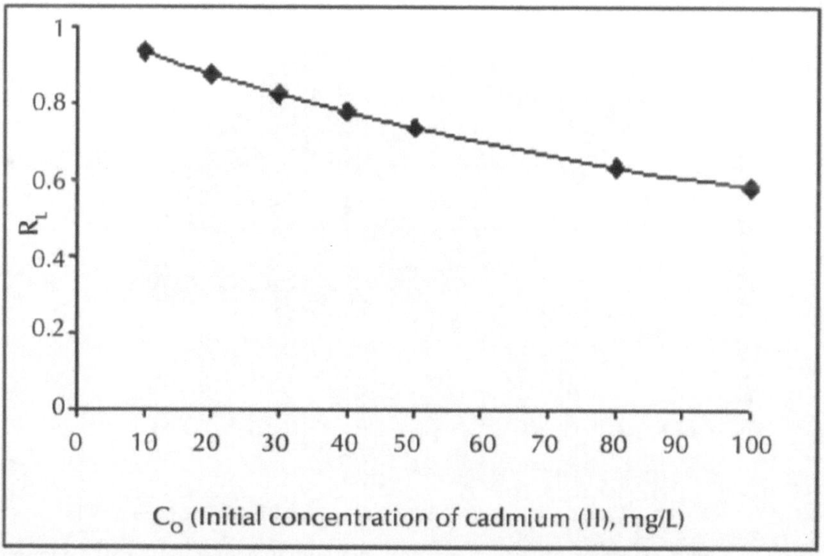

FIGURE 8 Effect of initial concentration of cadmium (II) on separation factor R_L.

9.3.3.3 Determination of Optimum Condition for Removal of Cadmium (II) by AIP Using Response Surface Methodology

The RSM has been employed to optimize the removal condition of cadmium (II) by AIP. Experimental condition as designed by Design Expert software is shown in Table 1. Three input parameters *viz.* initial concentration of cadmium (II), weight of AIP and pH have been varied according to the design and the percentage removal obtained at the specified condition is also shown in Table 1. The ANOVA analysis reveals that the quadratic model is significant. The final equation in terms of coded factors is given below.

$$R = 94.24 - 3.22 \times A + 10.51 \times B + 8.64 \times C + 2.19 \times A \times B - 0.087 \times A \times C - 0.56 \times B \times C - 3.19 \times A^2 - 7.47 \times B^2 - 15.21 \times C^2 \quad (4)$$

where, R = Percentage removal of cadmium (II) (%), A = Initial concentration of cadmium (II) (mg/l), B = Weight of *AIP* (g), and C = pH.

Figure 9 shows the combined effect of weight of AIP and initial concentration of cadmium (II) on percentage removal of cadmium (II) at constant pH ($C = 7$). Percentage removal of cadmium (II) has been obsereved to be increased upto a certain initial concentration (35 mg/l) at constant weight of AIP. When initial concentration is increased beyond that value, percentage removal has been found to be decreased. For instance, the percentage removal has been decreased from 78.62 to 67.86% when initial concentration of cadmium (II) was increased from 20 to 50 mg/l at constant weight of 3 g of AIP. From the Figure 9, it can be stated that higher weight of AIP and lower initial concentration of cadmium (II) (less than 35 mg/l) favor removal process. It is also noted from the figure that an increase in weight of AIP has more influence on removal process than a decrease in initial concentration of cadmium (II).

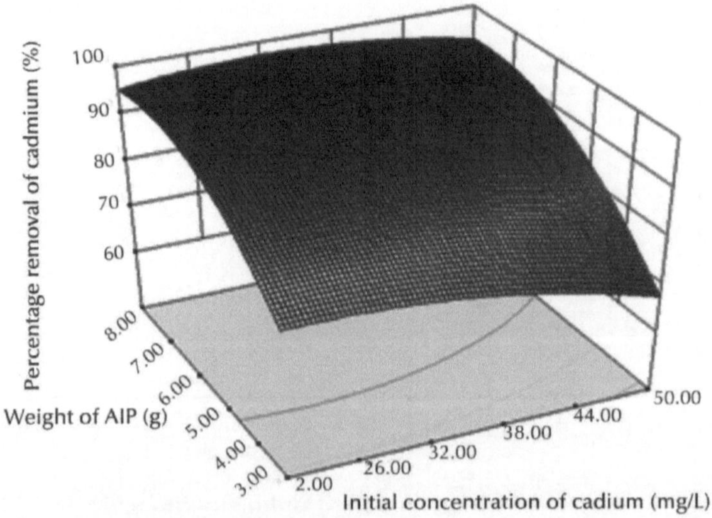

FIGURE 9 Combined effect of weight of AIP and initial concentration of cadmium on percentage removal of cadmium (II).

Figure 10 represents the combined effect of initial concentration of cadmium (II) and pH on removal of cadmium (II) at constant weight of AIP ($B = 5.5$ g). Percentge removal of cadmium (II) has been increased from 70.52 to 88.21% when pH increases from 4 to 10 at constant initial concentration of cadmium (II) 20 mg/l. However, increase in removal of cadmium (II) is less prominent due to increase in initial concentration of cadmium (II).

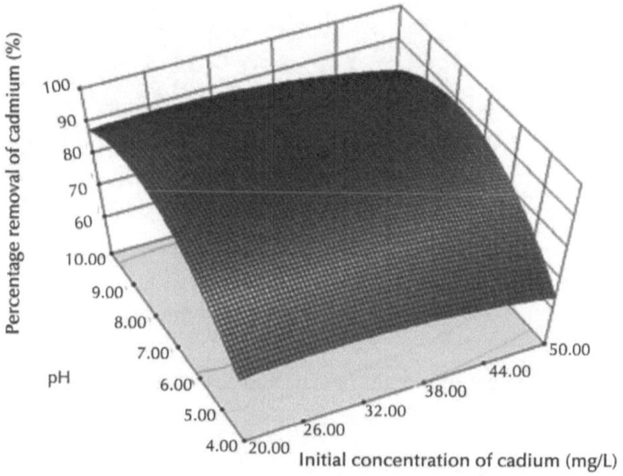

FIGURE 10 Combined effect of initial concentration of cadmium (II) and pH on percentage removal of cadmium (II).

Figure 11 embodies the combined effect of weight of AIP and pH on cadmium (II) removal at constant initial concentration of cadmium (II) (A=35 mg/l). Percentge removal of cadmium (II) has been increased from 52.44 to 70.69% and 74.14 to 90.76% when pH increases from 4 to 10 at constant weight of AIP 3 g and 8 g respectively. Therefore, it can be said that the percentage removal of cadmium (II) is strongly dependent on pH as well as weight of AIP, which is already observed in previous cases.

The actual and predicted responses are shown in Figure 12. Actual values have been obtained experimentally and predicted values have been obtained from the equation. The higher values of R^2 (0.9657) and R_{adj}^2 (0.9348) point out the good fitting of predicted response with the experimental one.

Optimization of removal process is aimed at making the process cost-effective. The following criteria have been set during optimization of removal process: initial concentration of cadmium (II): “in range”, weight of AIP: “equal to 5 g”, pH: “equal to 7” and response that is percentage removal: “maximize”. According to the software the optimum condition is obtained when 5 g of AIP is contacted with 26.41 mg/l cadmium (II) solution at pH 7 and 35°C temperature. Under the present optimized condition the predicted response (92.89%) matches well with the experimental observation (91.52%).

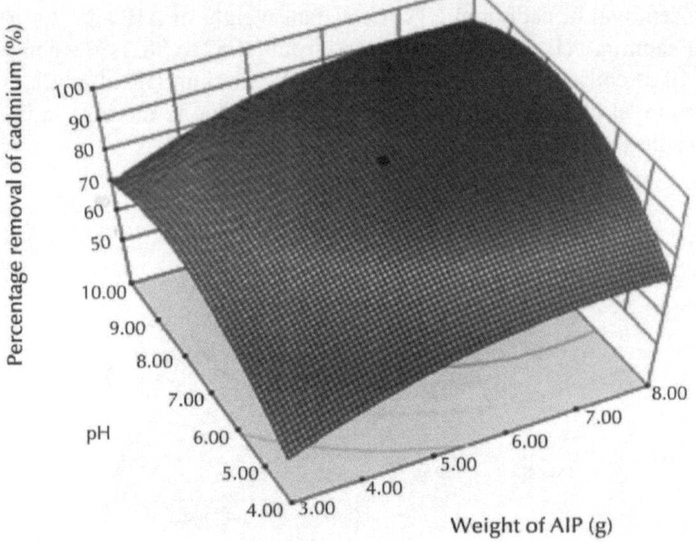

FIGURE 11 Combined effect of weight of AIP and pH on percentage removal of cadmium (II).

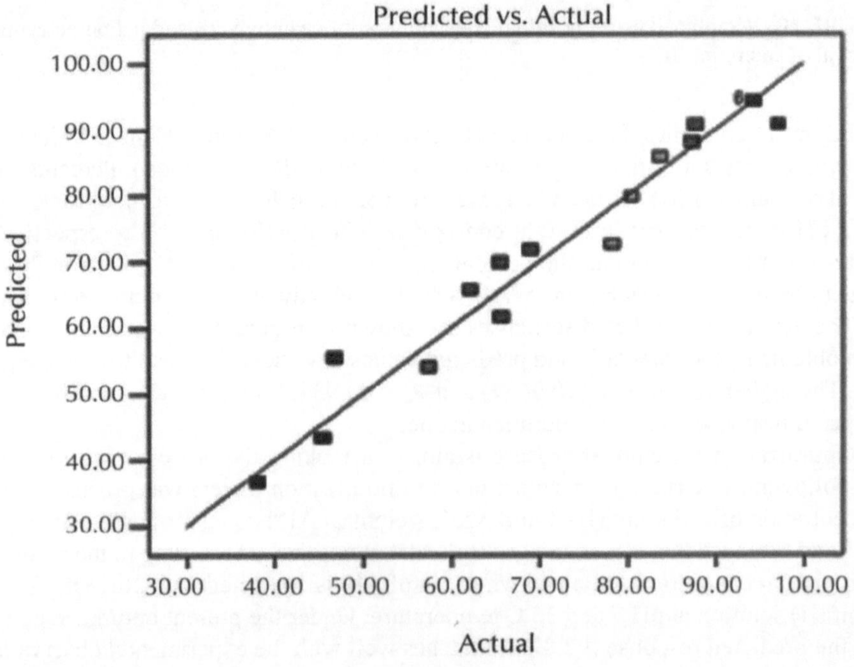

FIGURE 12 Plot of actual response versus predicted response for removal of cadmium (II).

9.3.4 Recovery of Cadmium (II)

Cadmium (II) recovery has been carried out by treating SAIP beads obtained after removal of cadmium (II) at optimum condition as specified by the software. The beads have been treated with acid as well as alkali solutions having different pH. Maximum 35.09% recovery of the adsorbed cadmium (II) is found at pH 4.0. Comparatively less recovery 21.13% and 9.9% was obtained at pH 7 and 9 respectively. It can be concluded that that papain immobilized in calcium alginate has been successfully used for recovery of cadmium (II) from simulated solution and may be used for recovery of cadmium (II) from industrial waste.

9.4 CONCLUSION

The present research work demonstrates adsorption of cadmium (II) from simulated solution using Papain enzyme. The enzyme immobilized in calcium alginate under specific condition designated as AIP was used as adsorbent. The removal condition is optimized using RSM. The optimized removal condition is: initial concentration of cadmium (II): 26.41 mg/l, weight of AIP: 5 g and pH: 7 at 35°C getting almost 92.88% removal of cadmium (II) from simulated solution. The cadmium (II) adsorption process is chemisorption in nature as the data fit most satisfactorily to PSOM.

Thereby, it can be concluded that AIP can efficiently be used for the treatment of waste water containing cadmium (II).

KEYWORDS

- **Adsorption**
- **Cadmium**
- **Kinetics and Equilibrium Study**
- **Recovery**
- **Response Surface Methodology**

REFERENCES

1. Holan, Z. R., Volesky, B., and Prasetyo, I. Biosorption of cadmium by biomass of marine algae. *Biotech Bioeng*, **41**, 819–825 (1993).
2. Sharma, Y. C. Economic treatment of cadmium (II)-rich hazardous waste by indigenous material. *J. Appl. Interface. Sci.*, **173**, 66–70 (1995).
3. Srinivasa Gowd, S. and Govil, P. K. Distribution of heavy metals in surface water of Ranipet industrial area in Tamil Nadu, India. *Environ Monit Assess*, **136**(1–3), 197–207 (2008).
4. Singh, R. *Urban impacts on groundwater quality in the Delhi region, Impacts of Urban Growth on Surface Water and Ground water Quality*. (Proceeding of IUGG 99 Symposium HSS, Birmingham, July, 1999). IAHS, 259 (1999).
5. Aktar, M. W., Paramasivam, M., Ganguly, M., Purkait, S., and Sengupta, D. Assessment and occurrence of various heavy metals in surface water of Ganga River around Kolkata: A study for toxicity and ecological impact. *Environ Monit Assess*, **160**, 207–213 (2010).
6. Malachowski, L., Stair, J. L., and Holcombe, J. A. Immobilized peptides/amino acids on solid supports for metal remediation. *Pure. Appl. Chem.*, **76**, 777–787 (2004).
7. Sluyterman, L. A. E. and Wijdenes, J. Proton equilibria in the binding of Zn^{2+} and of methylmercuric iodide to ppain. *Eur. J. Biochem*, **71**, 383–391 (1976).

8. Bhattacharyya, A., Dutta, S., De, P., Ray, P., and Basu, S. Removal of mercury (II) from aqueous solution using papain immobilized on alginate bead: Optimization of immobilization condition and modeling of removal study. *Bioresource Technology*, **101**, 9421–9428 (2010).
9. Al-Asheh, S., Banat, F., and Masad, A. Physical and chemical activation of pyrolyzed oil shale residue for the adsorption of phenol from aqueous solutions. *Environ. Geol*, **44**, 333–342 (2003).
10. Svilović, S., Rušić, D., and Bašić, A. Investigations of different kinetic models of copper ions sorption on zeolite 13X. *Desalination*, **259**, 71–75 (2010).
11. Dutta, S., Bhattacharyya, A., Ganguly, A., Gupta, S., and Basu, S. Application of response surface methodology for preparation of low-cost adsorbent from citrus fruit peel and for removal of methylene blue. *Desalination*, **275**, 26–31 (2011).

10 Biomass Based Integrated Power and Cooling Systems

T. Srinivas and B. V. Reddy

CONTENTS

10.1 INTRODUCTION

Biomass gasification involves the production of a gaseous fuel by partial oxidation of a solid fuel. Clean synthetic (syn) gas, produced from partial combustion of biomass can be burnt either in a gas turbine or a diesel engine combustion chamber to run a biomass based integrated energy system. By properly arranging the plant components, the simultaneous benefit of power and cooling can be obtained with the integrated system approach. In the current work, different options in integrated power systems and integrated combined power cooling systems are developed and discussed. Modeling and performance evaluation of such integrated energy systems have been addressed. In case of only power systems, a maximum of four power systems can be integrated as per the available heat recovery. In a combined power and cooling systems, maximum of three power systems can be integrated with one cooling system. It has been found that the combined power and cooling integration results high fuel efficiency compared to the only power integrated systems. The optimum compressor pressure ratio is decreasing with the integration of supplementary firing (SF) to the combined cycle power plant. Biomass has great potential as a renewable and clean energy for producing electricity, process heat and cooling with the integrated energy system approach. Gasification is a degradation process consisting of a sequence of thermal and thermochemical processes which converts the carbon in the solid fuel with restricted air into gases, leaving an inert residue. Biomass can be gasified in various ways by properly controlling the mix of fuel, air, and steam within the gasifier. The gasification of coal and biomass began in about 1800 and by about 1850 gas light for streets was common place. Due to its higher efficiency, it is desirable that gasification becomes increasingly applied in future rather than direct combustion. The current work is relevant because efficient conversion processes are required for renewable resources in order to compete with fossil fuels. Gasification of biomass is an attractive technology for combined heat, power and cooling production. Mark and Mike [1] discussed the use of biomass gasification process as the key element in an advanced gas turbine combined cycle system. Savola [2] simulated the possibilities to increase the power production and the power-to-heat ratio of 1–20 MW combined heat and power plants using biomass fuels with optimization tools. Margaret and Pamela [3] reported that the biomass systems produce very low levels of particulates, NO_x, and SO_x compared to the fossil systems. The biomass systems consume very small quantities of natural resources and have a positive net energy balance as they use renewable energy instead of non-renewable fossil fuels. Co-firing of biomass with coal offers us an opportunity to reduce the environmental burdens associated with the coal fired power systems. Anil et al. [4] solved the equations containing four atom balances (C, O, H, and N) and equilibrium relations for gas compositions using MATLAB at atmospheric conditions. Laihong et al. [5] simulated the processes, including chemical reactions and heat/mass balance with Aspen Plus software. Madhukar and Goswami [6] used a thermodynamic equilibrium model to predict the chemical composition of the products of biomass gasification. They carried out first law analysis for wood (designated by $CH_{1.5}O_{0.7}$) and showed that the optimum conditions for hydrogen production occurred at a gasification temperature of 1,000K, steam biomass ratio of 3 and equivalence ratio of 0.1. Krzysztof et al. [7] compared different types of biofuels for their gasification efficiency and evaluated for exergy efficiency. Rutherford [8] modeled biomass gasifier and studied the effects of steam fuel ratio and moisture content in biomass without taking solid carbon content in the synthetic (syn) gas. Although a lot of research and development work has been carried out during the past decade the commercial breakthrough for this technology is still to come. The detailed models of the integrated energy systems using biomass as a single fuel are not reported so far in the literature. The main objective of the current work is to present the new trends in the integrated power systems and integrated combined power and cooling systems. The current work is giving the conceptual suggestions in the combinations in the integrated energy systems.

10.2 INTEGRATED POWER SYSTEMS

Combined cycle is one of the good examples for the integrated power system. The merits of different individual systems can be integrated together to boost the performance and minimized the drawbacks and losses. In many cases it has been observed that there is a simultaneous demand for electric power and cooling at domestic and industrial level. To meet this demand, biomass has a great potential as a renewable and clean energy for combined heat and power (CHP) system. Recent fluctuations in natural gas fuel prices as well as environmental considerations have rekindled interest in alternatives to the natural gas based systems. Transforming solid fuels such as biomass into gas so that they can substitute for natural gas provides the opportunity to enhance the efficiency of biomass based power systems by allowing the solid fuels to be used in high efficiency power generation cycles such as integrated gasification combined cycle (IGCC).

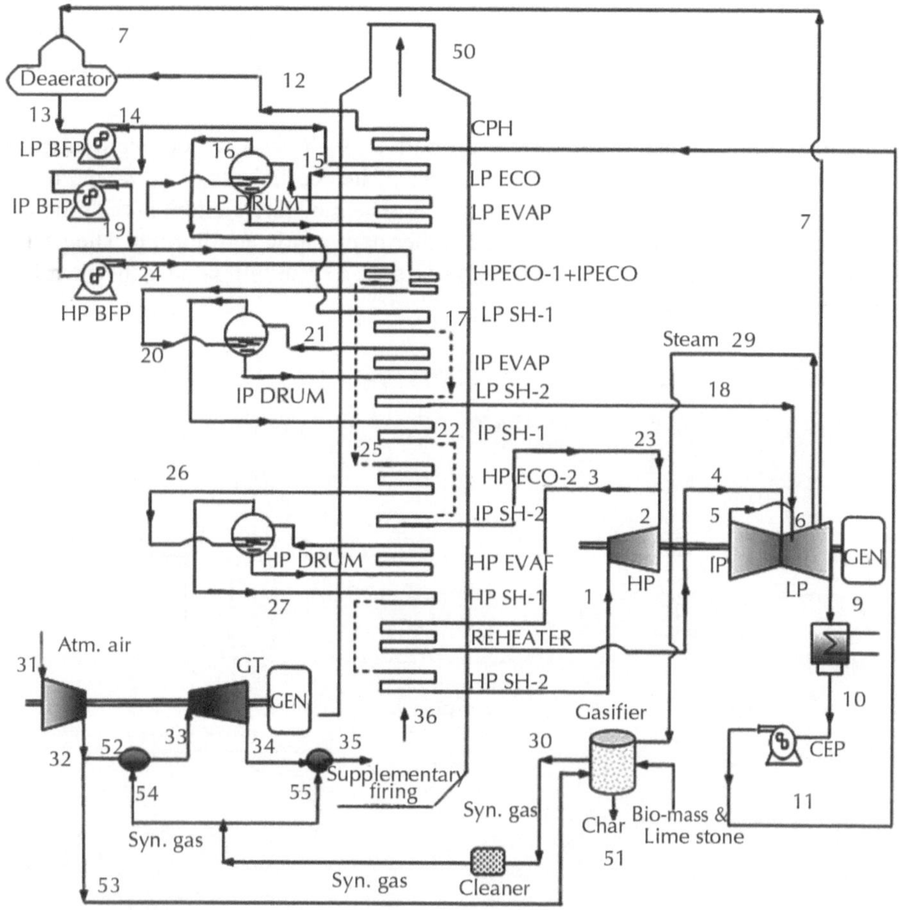

FIGURE 1 Biomass based integrated power system [9].

Figure 1 shows the schematic flow diagram of a biomass gasifier used in a combined power cycle. The biomass is fed at atmospheric conditions. The compressed air and steam enters the gasifier at 12 bar. The syngas after cleaning in a gas cleaner goes to gas turbine combustion chamber (GTCC) where a complete combustion takes place with supply of the compressed air. The exhaust gas from gas turbine generates steam in a heat recovery system from the feed water. The superheated steam from the heat recovery steam generator (HRSG) injected in the biomass gasifier and the remaining steam is expanded in a steam turbine. The syngas is assumed to be as an ideal gas and its properties are taken as a function of temperature. It is assumed that the steam enters into the gasifier at the pressure of the incoming compressed air. The heating devices in the HRSG are arranged to get the minimum temperature difference between the flue gas and the water/steam. The enthalpy rise between feed water inlet and steam outlet must equal the enthalpy drop of the exhaust gases in the HRSG, and the pinch point (PP) and terminal temperature difference (TTD) cannot be less than about 20°C if the HRSG is to be of economic size. The arrangement of heaters in the HRSG minimizes the exergetic loss in HRSG. Srinivas [10] proposed a new arrangement for the triple pressure heat recovery for a combined cycle power plant.

The biomass is defined by a general formula as $C_{a_1}H_{a_2}O_{a_3}N_{a_4}$. For single atom of carbon in fuel, the coefficient a_1 becomes to one. The coefficients of a_2, a_3 and a_4 are the H/C, O/C, and N/C mole ratios, respectively. The moisture content in the biomass fuel is neglected. The reactions are solved at thermodynamic equilibrium. The gasification products contain CH_4, CO, CO_2, H_2, H_2O, and N_2. The following is the chemical reaction in biomass gasifier [11].

$$C_{a_1}H_{a_2}O_{a_3}N_{a_4} + a_5(O_2 + 3.76N_2) + a_6H_2O$$
$$\Rightarrow b_1CH_4 + b_2CO + b_3CO_2 + b_4H_2 + b_5H_2O + b_6N_2$$

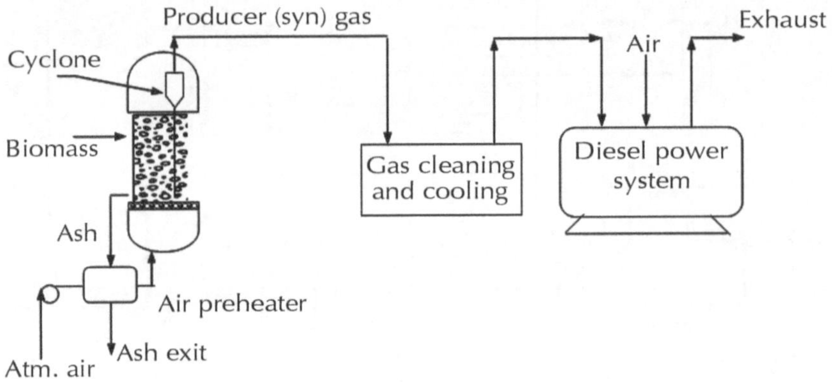

FIGURE 2 Biomass based diesel power system.

Figure 2 shows the biomass based integrated power system using a diesel power plant. In this plant, the producer gas coming from the gasifier is cooled and filtered before the induction to diesel engine. Here a lot of heat loss takes place, nearly from 550°C to 35°C. There is a scope to recover the heat from the producer gas as well as from the diesel engine exhaust. There is one more option of running this diesel engine with the dual fuel that is diesel and producer gas.

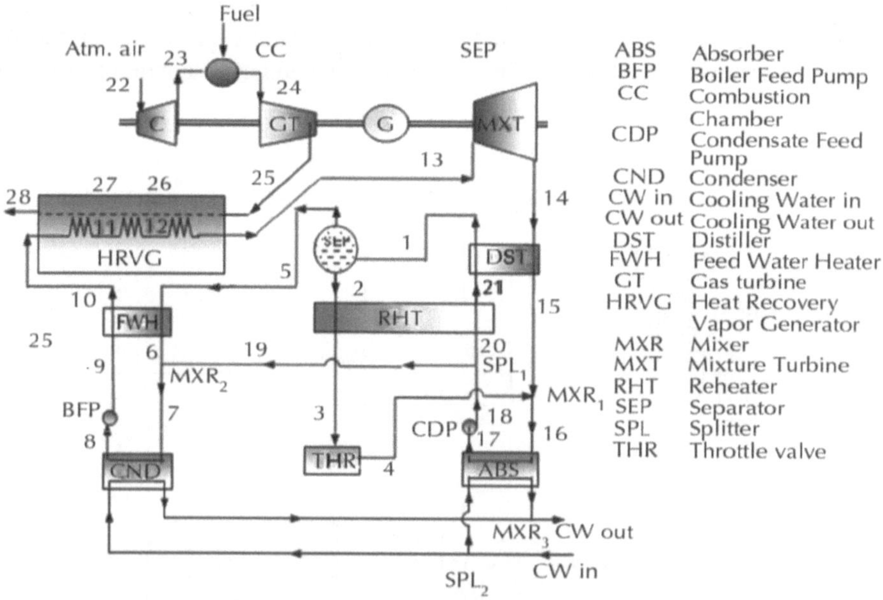

FIGURE 3 Integration of high temperature based Kalina power system to the gas power plant in place of steam bottoming plant [12].

The schematic flow diagram of the combined cycle with simple Brayton cycle as a topping one and coupled with the Kalina cycle, is shown in Figure 3. The ammonia mixture turbine exhaust (14) of Kalina cycle is cooled in the distiller, then diluted with a weak solution (4) and condensed in the absorber. The saturated liquid leaving the absorber is heated in a distiller by the turbine exhaust. The working fluid is separated into rich ammonia water vapor (5) and poor liquid mixture (2) in the separator. The rich vapor mixture is mixed with the condensate from the splitter to obtain the desired concentration. It is condensed in a condenser, compressed, heated in a feed water heater and sent to the boiler, where it is super heated by the exhaust of the gas turbine. The superheated vapor expanded in the turbine and gives output from the Kalina cycle. Rich ammonia vapor from separator cools in feed water heater. In the separator, both the liquid and vapor quantities are separated.

Figure 4 shows the schematic flow diagram of a biomass based integrated power plants with the gas, steam, and aqua ammonia flows and a typical HRSG. The producer gas temperature is around 550–600°C, where there is a chance to recover the heat from the hot producer gas. A high temperature Kalina power plant can be operated using this heat recovery. The working of high temperature Kalina cycle has been reported in the previous section. The syngas after the heat recovery and cleaning in a gas cleaner goes to GTCC where a complete combustion takes place with supply of the compressed air. The flue gas enters the HRSG and is reduced in temperature by the reheater, superheater, drum evaporative surfaces, and economizer before it enters the stack. The exhaust of HRSG can be used to generate ammonia-water vapor mixture in heat recovery vapor generator (HRVG) to generate power after expansion of this mixture. The temperature of vapor is raised in super heater before its expansion. Shankar Ganesh and Srinivas [13] developed the optimum plant and cycle conditions for a low temperature based Kalina power plant.

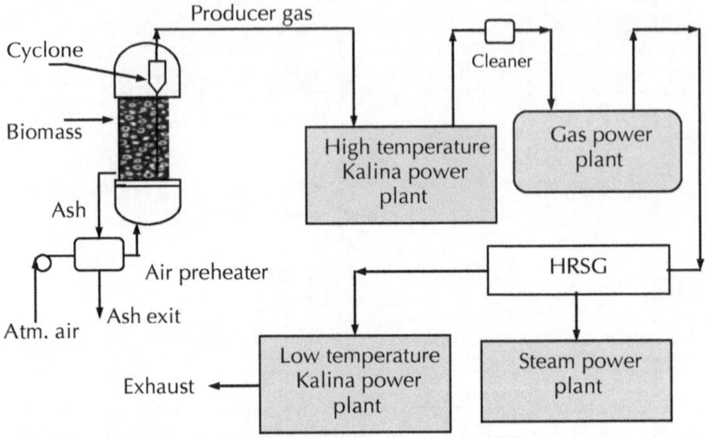

FIGURE 4 Integrated power systems consists of high temperature Kalina power plant, gas power plant, steam power plant, and low temperature Kalina power plant.

10.3 INTEGRATED POWER AND COOLING SYSTEMS

The combined power and cooling can be obtained from a single fuel plant using the integrated approach of combined power and cooling systems.

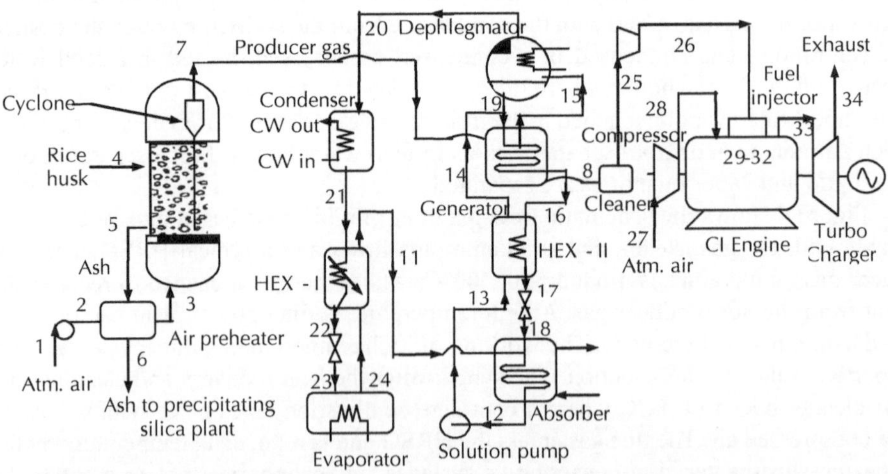

FIGURE 5 Schematic process flow diagram of integrated biomass gasification, cooling system, and diesel power system.

Figure 5 shows the processes details of biomass gasification, power generation and refrigeration in the integrated plant. The atmospheric air from the blower enters the gasifier where the pressure is maintained slightly above the atmospheric pressure. The syngas after rejecting heat in vapor generator and cleaning in a gas cleaner goes to diesel engine where a complete combustion takes place with supply of the compressed air. The diesel engine is a turbocharged model. In VAR system, the function of rectification column and dephlegmator is to reduce the concentration of water vapor at the exit of the generator. Without these the vapor leaving the generator may consist of five to ten percent of water. However, with rectification column and dephlegmator the concentration of water is reduced to less than one percent. An almost pure ammonia leaves the evaporator, exchanges heat with the condensed liquid in HEX-I and enters the absorber. This refrigerant is absorbed by the weak solution coming from the solution expansion valve. The heat of absorption is rejected to an external heat sink. Next the strong solution that is now rich in ammonia leaves the absorber and is pumped by the solution pump to generator pressure. This high pressure solution is preheated in HEX-II and then heated in the generator with syngas heat recovery. The condensed liquid is subcooled in the subcooling heat exchanger (HEX-I) by rejecting heat to the low temperature, low pressure vapor coming from the evaporator. The subcooled, high pressure liquid is then throttled in the refrigerant expansion valve. The low temperature, low pressure and low quality refrigerant then enters the evaporator, extracts heat from the refrigerated space and leaves the evaporator. From here it enters the subcooling heat exchanger to complete the refrigerant cycle. Now, the condensed water in the dephlegmator flows down into the rectifying column along with rich solution and exchanges heat and mass with the vapor moving upwards. The hot solution that is now weak in refrigerant flows into the solution heat exchanger where it is cooled by preheating the rich solution. The weak but high pressure solution is then throttled in the solution expansion valve, from where it enters the absorber to complete its cycle (HEX-II).

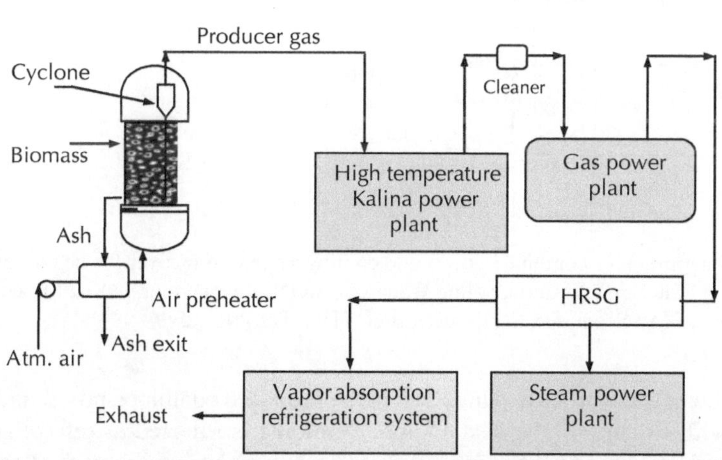

FIGURE 6 Integration of three power plants and one cooling system using biomass as a fuel [14].

Figure 6 shows the integration of biomass gasifier, high temperature Kalina power plant, gas power plant, steam power plant and vapor absorption refrigeration system. Compressor supplies air to GTCC and gasifier at the same pressure. The resulted hydrogen rich syngas is used as high temperature fuel source to Kalina power plant. Heat is recovered in the ammonia vapor generator and the low temperature producer gas is cooled in a cleaned. A steam power plant powered by a triple pressure heat recovery generates the power in addition to gas plant and Kalina plant. Still there is a scope to recover the heat from the exhaust at the exit of HRSG. This can be recovered by integrating a single effect heat operated cooling system. The addition of this cooling plant to the IGCC plant increases the energy utilization which limits the global warming. The VAR system works on the principle of separation (generator) and mixing (absorber) of aqua ammonia mixture at high and low temperature respectively. Ammonia is selected as refrigerant and water as absorbent. Due to difference in their boiling points ammonia first starts to evaporate in generator from the heat recovery of exhaust gas. So, this integrated system supplies power from the three plants that is gas, steam, and Kalina plants and cooling from VAR system.

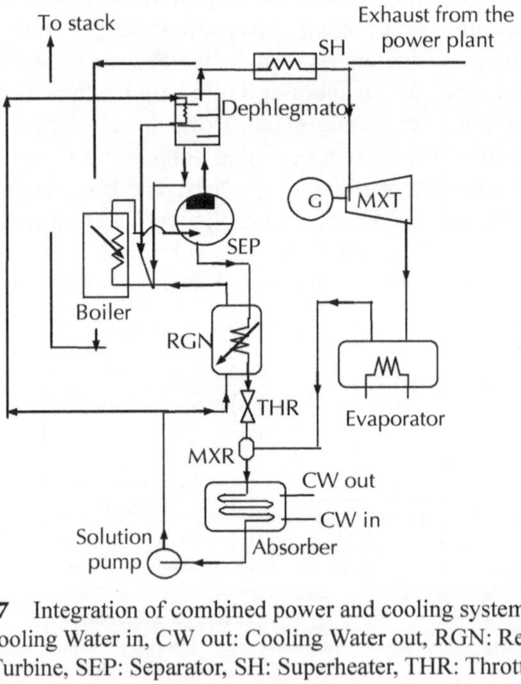

FIGURE 7 Integration of combined power and cooling system to recover the heat of exhaust, CW in: Cooling Water in, CW out: Cooling Water out, RGN: Regenerator, MXR: Mixer, MXT: Mixture Turbine, SEP: Separator, SH: Superheater, THR: Throttle valve.

Figure 7 depicts the processes involved in the proposed combined power and cooling system with exhaust as the heat source. Ammonia is selected as refrigerant and water as absorbent. Due to difference in their boiling points ammonia first starts to evaporate in vapor generator from the heat recovery of the exhaust gas. The heat from the exhaust coming from the HRSG of combined cycle power plant is recovered in vapor generator (boiler). In separator, the working fluid is separated into rich ammonia water vapor and weak liquid mixture. Still, some moisture content in the rich mixture presents due to high boiling point of water compared to ammonia. The traces of moisture in ammonia vapor are separated by cooling in dephlegmator. After superheating, the ammonia vapor is expanded in the mixture turbine. At the exit of this turbine, the expanded fluid temperature is below the atmospheric temperature. It absorbs the heat in the evaporator where the vopor temperature increases and produces cooling. The weak solution coming from throttle and the evaporator exit fluid mixed together and the mixture heat is rejected at the absorber to get the saturated liquid condition.

The schematic diagram of a simple gas cycle integrated with HRSG, steam plant and VAR is shown in the Figure 8. Topping power plant is GTs plant where as the bottoming plant is the steam turbine based plant. The two plants are linked with a dual pressure HRSG. The heating device in HRSG are arranged to get the minimum temperature difference between gas and water/steam. Single effect LiBr/H_2O based VAR cooling system is proposed to cool the compressor inlet air cooling. The cooling system evaporator is coupled to the air cooler to decrease the atmospheric air temperature. A chilled water circuit is located between these two heat exchangers that is evaporator and air cooler to transfer heat from cooling coil to air. The VAR cooling system is power by the heat recovery at plant stack. This serves as a low grade thermal energy heat source for LiBr/H_2O system supplying about 80–130°C. Temperature at the inlet of the GT plant is taken as ambient temperature about 30°C for normal (simple gas cycle) and for air temperature about 14°C with VAR cycle.

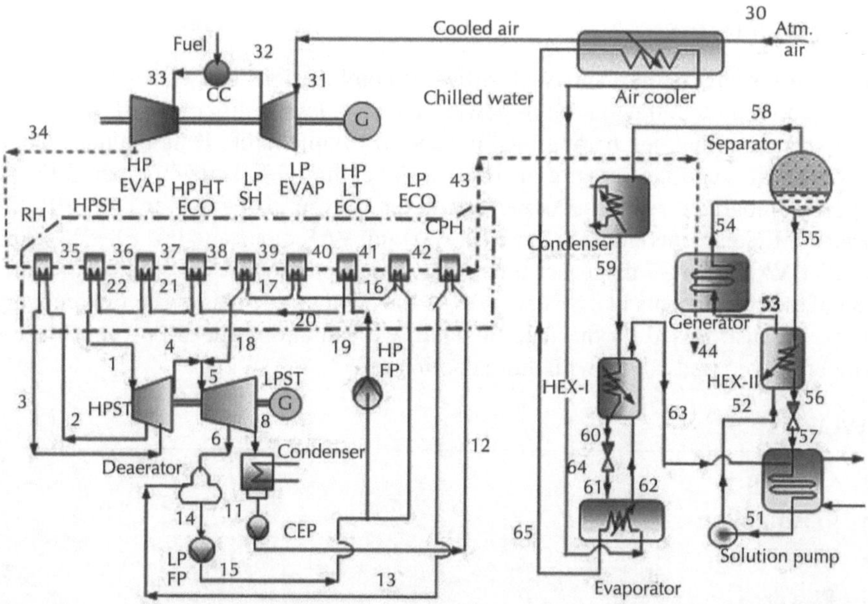

FIGURE 8 Integrated cooling system to a power plant for compressor inlet air cooling.

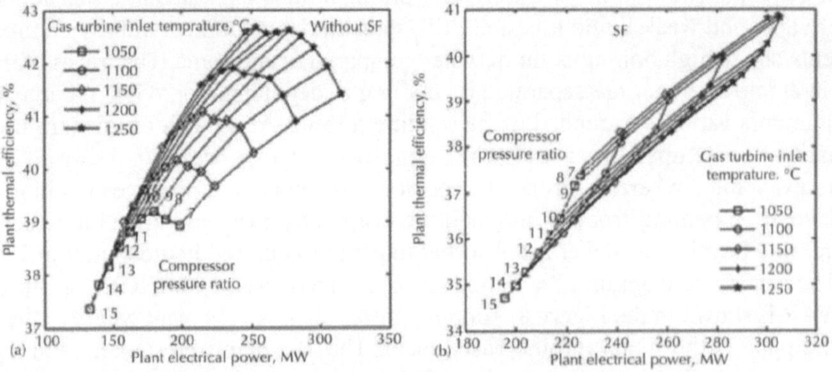

FIGURE 9 Performance details of IGCC power plant with and without SF system.

Figure 9 presents the performance characteristics of the IGCC plant with compressor pressure ratio and gas turbine inlet temperature (a) without SF and (b) with SF. It has been found that there is an optimum pressure ratio for each set of turbine inlet temperature without SF and this optimum value increases with temperature. The optimum pressure ratio decreases with addition of SF. Thus introduction of SF demands comparatively low compressor pressure ratio. The optimum pressure ratios without SF are 9, 10, 11, 12, and 14 respectively for the gas turbine inlet temperatures of 1,050, 1,100, 1,150, 1,200, and 1,250°C. For SF plant, these optimum values are shifting towards the lower value side.

10.4 CONCLUSION

Biomass based integrated plants with different combinations and configurations are proposed and reported. Integration of power systems allows a maximum of four plants that is gas, steam, high temperature Kalina and low temperature Kalina plants. Combined power and cooling integration results higher fuel efficiency compared to the only power integrated systems. As per the literature results of Srinivas et al. [9], the integrated efficiency increment is more (0.9%) with VAR alone and less (0.25%) from SF to SF + VAR. The SF decreases the optimum compressor pressure ratio. The integration of the components like SF and VAR to the power system allows more recovery from the exhaust. Based on the load demand, we can choose peak power with low cooling effect or rated power with more cooling effect.

KEYWORDS

- **Cooling**
- **Gasification**
- **Integration**
- **Power**
- **Thermodynamics**

REFERENCES

1. Mark, A. and Mike, J. W. *Biomass gasification combined cycle opportunities using the future energy silvagas gasifier coupled to Alstom's industrial gas turbines*. Proceedings of ASME Turbo Expo 2003, Georgia World Congress Center, No.GT2003-38294, pp. 1–7 (2003).
2. Savola, T. *Simulation and optimization of power production in biomass-fuelled small-scale CHP plants*. M. S. Thesis, Helsinki University of Technology Department of Mechanical Engineering Energy Engineering and Environmental Protection Publications, Espoo 2005, TKK-ENY-23 (2005).
3. Margaret, K. M. and Pamela, L. S. *Life cycle assessment comparisons of electricity from biomass, coal, and natural gas*. Paper No.18d, National Renewable Energy Laboratory, Annual Meeting of the American Institute of Chemical Engineers (2002).
4. Anil, K., Prasad, P., Preeti, A., and Anuradda, G. *Equilibrium model for biomass gasification*. Proceedings of International conference on Advances in Energy Research (AER–2006), pp. 106–112 (2006).
5. Laihong, S., Yang, G., and Jun, X. Simulation of hydrogen production from biomass gasification in interconnected fluidized beds. *Biomass and Bioenergy*, **32**(2), 120–127 (2008).
6. Madhukar, R. M. and Goswami, D. Y. Thermodynamic optimization of biomass gasifier for hydrogen production. *International Journal of Hydrogen Energy*, **32**(16), 3831–3840 (2007).
7. Krzysztof, J. P., Mark, J. P., and Anke, P. Exergetic evaluation of biomass gasification. *Energy* **32**(4), 568–574 (2007).
8. Rutherford, J. *Heat and power applications of advanced biomass gasifiers in New Zealand's wood industry*. M. E. thesis in chemical and process engineering, University of Canturbury (2006).
9. Srinivas, T., Reddy, B. V., and Gupta, A. V. S. S. K. S. Biomass fueled integrated power and refrigeration system. *Journal of Power and Energy, Proceedings of the Institution of Mechanical Engineers Part A, Professional Engineering Publishing*, **225**(3), 249–258 (2011a).
10. Srinivas, T. Study of a deaerator location in triple pressure-reheat combined power cycle. *Energy*, **34**(9), 1364–1371 (2009).
11. Srinivas, T. Gupta, A. V. S. S. K. S., and Reddy, B. V. Thermodynamic equilibrium model and exergy analysis of a biomass gasifier. *ASME Journal of Energy Resources Technology* **131**(3), 1–7 (2009).
12. Srinivas, T. Gupta, A. V. S. S. K. S., and Reddy, B. V. Performance simulation of combined cycle with Kalina bottoming cycle. *Cogeneration and Distributed Generation Journal, the Association of Energy Engineers Press, Taylor & Francis Group*., **23**(1), 6–20 (2008).
13. Shankar Ganesh, N. and Srinivas, T. Design and modeling of low temperature solar thermal power station. *Applied Energy*, **91**(1), 180–186 (2011).
14. Srinivas, T. Reddy, B. V., and Gupta, A. V. S. S. K. S. *Integration of Hybrid power system using biomass fuel*. International Conference on Harnessing Technology, February 13th and 14th, 2011, Caledonian College of Engineering, Muscat (2011b).

11 Reclamation of Wastes for Mercury Removal —A Review

Aparupa Bhattacharyya, Srabanti Basu, and Susmita Dutta

CONTENTS

11.1 INTRODUCTION

Mercury pollution has turned one of the most severe environmental problems due to its persistence in nature and toxicity to all forms of life. Many techniques exist to remove mercury but such processes have their own demerits. To find out an effective technology for removal of mercury to the desired level is now the main concern of the researchers. Among the various existing effluent treatment technologies, adsorption process is the most popular and effective one practiced throughout the world and activated carbon has been proven as the most suitable adsorbent. But the use of activated carbon is restricted due its high cost. Nowadays researches are going on to find out economical as well as efficient adsorbent prepared from waste materials. This chapter presents several investigations carried out to prepare adsorbent from different types of waste materials to remove mercury from effluent as well as from drinking water. Though quite a few adsorbent have shown high adsorptive capacity still some of them have a number of shortcomings. It is evident from the review that many waste materials have scope to be an effective precursor for preparation of adsorbent for removal of mercury.

Heavy metals are elements having atomic weight between 63.5 and 200.6 and a specific gravity greater than 5. They are introduced in the environment through the discharge of untreated or improperly treated industrial waste. Due to their non-biodegradable, recirculating, and persistence characteristics, they are bioaccumulated in the ecosystem causing severe environmental pollution problem. Mercury (atomic weight: 200.59 and specific gravity: 13.6) is one of the most threatening contaminants among those heavy metals. It ranks third in the list of priority pollutants prepared by Agency of Toxic Substances and Disease Registry (ATSDR). Mercury pollution is a global problem due to its wide distribution in nature and its toxicity to all forms of life ranging from bacteria to higher eucaryotes like plants and mammals. The total global input of mercury has been estimated to be 10^{10} g. About 10–50% of this flux originates from human activities [1]. The industries discharging mercury contaminated wastewaters include chloro alkali, smelting, tars, and asphalt, coke ovens, textiles and those manufacturing cements, catalysts, paints, pesticides, pharmaceuticals, and batteries [1]. Thus, proper treatment of mercury laden effluents is necessary from environmental pollution point of view. As per World Health Organization (WHO), the maximum permissible limit in drinking water is 0.001 ppm [2]. Although several conventional techniques like chemical precipitation, chemical coagulation, ion exchange, electrochemical method, reverse osmosis, and so on are being used to treat the mercury laden effluent, each of the processes has its own limitation. When chemical precipitation process is associated with the problem of solid disposal, the chemical coagulation, ion exchange, reverse osmosis, and so on cannot reduce mercury levels to meet the standard limit as prescribed by different environmental agencies [2, 3]. Furthermore, ion exchange, reverse osmosis process, and so on demand high capital cost which is prohibitive for medium and small scale industries [2]. Among these various technologies, adsorption being technologically simple, cost effective, and environmental friendly can be considered as better alternative for removal of mercury from waste effluent. The term adsorption refers to the accumulation of a substance at the interface between two phases such as solid and liquid or solid and gas [4]. The substance that accumulates at the interface is called "adsorbate" and the solid on which adsorption occurs is called "adsorbent" [4]. Though several commercial adsorbents such as silica gel, zeolites, activated alumina and so on are in practice to remove environmental contaminants from waste effluents, activated carbon plays a major role in removal pollutants due to its microporous structure, vast surface area and high adsorption capability [4, 5]. However, high capital and regeneration cost of the commercial activated carbon are the main drawbacks which restrict its widespread use in industrial scale. Thus, researches are now tending towards the development of new low cost adsorbent from various non-conventional waste materials from industries and agriculture, having equivalent potential as that of commercial adsorbent [5]. In general, an adsorbent is considered as low cost when it requires little processing during preparation from abundantly available natural product or agricultural/industrial wastes [6]. The present chapter depicts an overview on the removal of mercury using low cost adsorbents prepared from various non-conventional waste materials. Though a number of review papers are published in this line, a comprehensive study on the adsorption of mercury using reclaimed waste material was in need. Thus, in the present paper an attempt has been made to give an overview of the studies based on the abatement of mercury using low cost adsorbent prepared from waste material. Furthermore, studies presenting the recovery of mercury and subsequent reuse of adsorbent are also mentioned in the present chapter.

11.2 ADSORBENT FROM DIFFERENT WASTES FOR REMOVAL OF MERCURY

The classification of waste materials used for preparation of low cost adsorbent is represented in Table 1.

TABLE 1 Classification of waste materials to be used as precursor for preparation of adsorbent.

Types of wastes			
Agricultural	**Animal**	**Industrial**	**Miscellaneous**
Rice husk	Algerian sheep hooves	Rice husk ash	Eucalyptus wood
Coconut husk	Crab carapace	Fly ash	Walnut shell
Ceiba pentandra hulls	Camel bone	Tire granules	Bamboo
Phaseolus aureus hulls		Sago industry wastes	Wood powder from suji
Cicer arietinum waste		Palm oil empty fruit bunch	Olive stones
Coconut coir pith			Date palm leaflet
Cotton seed			Guava bark
Tamarind nut			
Corn processing waste water			
Starch			
Bagasse pith			

11.3 AGRICULTURAL WASTES

11.3.1 Husk or Hulls of Fruits, and Grains

Rice is one of the major crops grown all over the world. It is the most important cereal grain which is used as staple food for a large part of human population. Barley is very nutritious cereal grain. Coconut is also a very popular fruit, rich in fiber, vitamins, and minerals. Large production of rice, coconut, barley, and so on generates tons of husks as waste. Many researches are going on throughout the world to utilize these waste materials as an adsorbent for removal of various pollutants.

The effectiveness of rice husk, a lignocellulosic waste, has also been scrutinized by Krisnani et al. [7], when a biomatrix prepared from rice husk after alkali treatment was used for removal of Hg (II). Effect of pH and initial metal concentration on adsorption capacity were studied. The majority of metal ion was removed within first 90–120 min and equilibrium was achieved within 120–150 min of contact time. The maximum adsorption capacity was found to be 33.1 mg/g of adsorbent when metal concentration was 200 mg/l. Partial alkali digestion increases the surface area that facilitated the transport of metal ion to the binding site. Rice husk contains enough calcium, magnesium and -OH, -COOH groups in the lignocellulosic moieties. These groups served as the binding sites for Hg (II). The research revealed that metal binding is strongly dependent on pH with more metal cations bound at higher pH and the maximum uptake of metal ion took place at pH 5.5 – 6 ± 0.1. The equilibrium data fitted well to the Langmuir adsorption isotherm model. To assess the effectiveness of the process, column study was done in addition to batch adsorption. Research revealed that there was no leakage up to 25–33 bed volumes with an initial concentration of 12.5 mg/l and the maximum removal of metal ions was achieved at flow rate 0.2 ml/min.

In another study, a carbonaceous sorbent was prepared from rice husk via sulfuric acid treatment by El-Shafey [8]. Rice husk was carbonized using sulfuric acid. The maximum Hg (II) uptake took place in between pH 5–6. Metal uptake increased with temperature. Isotherm data fitted well to the Langmuir equation. The optimum adsorption capacity of Hg (II) (384.62 mg/g) was obtained with wet adsorbent at 45°C whereas the adsorption capacity decreased to 303.03 mg/g at 45°C when dry adsorbent was used. Study revealed that sorption of Hg (II) followed pseudo-second order kinetic model and the activation energy was found to be 54 kJ/mol implying the chemical reaction controlled process. Shafey [9] proposed the following mechanism for sorption of Hg(II) on the prepared adsorbent from chloride media at different pH values. As per stability constant calculations in the presence of chloride ion at pH less than 4, the principal component is $HgCl_2$. As at low pH, chloride anions have a propensity for formation of more stable complexes with mercury such as $HgCl_2$, $HgCl_3^-$,s and $HgCl_4^{2-}$, it should inhibit the adsorption of Hg (II) onto the sorbent at that condition. Moreover, at low pH condition, the presence of excess H^+ ions in solution hinders the sorption of Hg (II) by competing with Hg (II) ions for active sites. Shafey [9] also presumed the reduction of Hg (II) to Hg (I) as evidenced by the presence of Hg (I) chloride crystals on the sorbent surface. Hg (II) reduction is represented as follows:

$$2M^{2+} + 2e = M_2^{2+} \qquad E_0 = +0.92V$$

The reason behind the increased sorbent acidity and cation exchange capacity might be due to the occurrence of the following reaction as proposed by El-Shafey [8].

$$\sim C-H+M^{2+}+H_2O=C-OH+M_2^{2+}+H^+\sim C-OH+M_2^{2+}+H^+$$

$$\sim C-H/\sim C-OH+M^{2+}+H_2O=\sim CO+M_2^{2+}+H^+$$

$$\sim C-H/\sim C-OH+M^{2+}+H_2O=\sim COOH+M_2^{2+}+H^+$$

Where '*M*' denotes '*Hg*'.

Hsi et al. [10] (2011) used barley husk and rice husk as an adsorbent to remove elemental mercury (Hg^0) and showed the better efficacy of agricultural residue based adsorbent in Hg^0 removal than that derived from coal. Two agricultural residues, including biotreated barley husk and rice husk, were chosen for producing carbonaceous adsorbents. Samples were prepared by carbonization at 400°C and steam activation at 800°C. To incorporate sulfur in the adsorbent, samples were mixed with elemental sulfur at high temperature. Ultra-high purity N_2 flowed continuously to maintain an oxygen-free environment. Sulfur impregnation is done not only to biotreated barley husk and rice husk but also to commercial activated carbon to compare their adsorption capability. Due to the microporous structure, the adsorbent without any sulfur treatment showed the ability to uptake Hg^0. The Hg^0 binding capacity was attributed due to the presence of functional groups like carbonyl and carboxyl groups. In sulfur treated adsorbent, oxygen and sulfur functional groups accounted for the Hg^0 adsorption. The adsorption capacities of biotreated barley husk and rice husk were initially compared with commercial activated carbon in its original form. Without any sulfur treatment the biotreated barley husk and rice husk showed the equilibrium adsorption capacity as 165.8 and 161.2 μg/g whereas the equilibrium adsorption capacity for commercial activated carbon was found to be 119.3 μg/g without sulfur treatment. Study revealed that the adsorption capacities of the samples were increased due to sulfur impregnation. Sulfur treated all the samples showed the equilibrium capacity between 119.2 and 344.3 μg/g. The sulfur treated rice husk at 650°C gave the maximum adsorption capacity as 344.3 μg/g. The Hg^0 adsorption kinetics was best described by the pseudo second order model implying the chemisorptions nature of reaction and Langmuir adsorption isotherm model was found to fit the equilibrium data. The involvement of two active sites for binding of one molecule of Hg^0 had been suggested in the present chapter.

To explore the possibility of coconut husk, an agricultural waste, Hasany and Ahmad [11] have used it for removal of Hg (II) from aqueous media. Dried coconut husk was ground to a particular size (0.42 mm) and stored for further use. Sorption of Hg (II) was examined under varying nature of electrolyte, amount of coconut husk, and equilibration time. Study showed that adsorption of Hg (II) on coconut husk was influenced by pH of solution and depended on nature of electrolyte present in solution. When deionized water was used as an electrolyte in the sorption medium, the removal of Hg (II) was 90.9%. Sorption of metal ion attained equilibrium within 20 min.

Amount of adsorbent (10–300 mg/4.5 ml) affected the adsorption process directly. The percentage sorption increased when amount of adsorbent was raised and became indifferent above 100 mg/4.5 ml. The equilibrium data of Hg (II) adsorption were fitted best to the Freundlich isotherm model. The negative values of ΔH and ΔG, indicated that the process is endothermic and spontaneous. Study also showed that adsorption of Hg (II) was significantly reduced in presence of different anions and cations. Selectivity test disclosed that the adsorbent is selective to different metal ions. The significant uptake of Hg (II) onto coconut husk may be attributed to several hydroxyl and carboxylic groups present in the adsorbent matrix.

Different types of agricultural residue have been tested by Rao et al. [12] to investigate their Hg (II) removal capability under different parameters like pH, contact time, initial metal concentration and adsorbent amount. Activated carbon derived from three types of agricultural wastes and by-product namely *Ceiba pentandra* hulls (ACCPH), *Phaseolus aureus* hulls (ACPAH) and *Cicer arietinum* waste (ACCAW) were used to adsorb Hg (II) from waste water. The hulls were carbonized and steam activated at an elevated temperature. Among the three adsorbents, ACCPH is found to be the most potential one. The uptake of Hg (II) increased with increase in pH of aqueous solution and attained maximum value at pH 6.0 in the case of ACCPH. On the other hand, for ACPAH and ACCAW, the maximum Hg (II) removal was observed at pH 7.0. More than 50% of Hg (II) was adsorbed within 10 min and equilibrium adsorption was attained in 90, 100 and 110 min for the adsorbents ACCPH, ACPAH and ACCAW respectively. Study showed that removal capacity of all the adsorbents increased with increase in initial concentration. The removal of Hg (II) increased from 2.5 to 29.20 mg/g for ACCPH, from 2.5 to 23.37 mg/g for ACPAH and from 2.5 to 21.49 mg/g for ACCAW when initial concentration of Hg (II) was increased from 10 to 150 mg/l, respectively. It was also observed that with increase in adsorbent dosage from 25 mg to 300 mg, the percent removal of Hg (II) increased up to 99.7, 98.0, and 96.29% for ACCPH, ACPAH, and ACCAW respectively. Further increase in dosage has no effect on Hg (II) removal. Maximum removal was reported to be 29.20 mg/g of ACCPH when initial concentration of Hg (II) was 150 mg/l. The adsorption equilibrium data was best described by the Freundlich adsorption isotherm model and the kinetics of adsorption followed the pseudo second order model. The ACCPH was found to be the best adsorbent compared to other two. Higher adsorption of Hg (II) onto ACCPH adsorbent may be attributed to the presence of surface sulfur groups.

11.3.2 Coir and Bagasse Pith

Coir pith, an abundant lignocellulosic agricultural solid waste, was recycled and reused to develop $ZnCl_2$ activated carbon by Namasivayam and Sangeetha [13] for treatment of waste water containing different pollutants including heavy metals. Coir pith was chemically activated with zinc chloride. For an initial concentration of 20 mg/l of Hg (II), 100% removal was obtained with prepared $ZnCl_2$ activated coir pith carbon whereas 80% removal was obtained with coir pith carbon.

Coconut coir pith, an unwanted agricultural solid by-product, was used by Namasivayam and Kadirvelu [14] to assess its potentiality for removal of Hg (II) from waste water by varying agitation time, initial metal concentration, adsorbent dosage and pH. Coirpith was carbonized with sulfuric acid and ammonium persulphate at an elevated temperature. Experiment disclosed that coirpith carbon would require less residence time for the complete removal of Hg (II) than that for pea nut hull carbon and commercial activated carbon. The percentage removal of Hg (II) increased with increase in adsorbent dosage. It also increased with the increase of pH from 2 to 5 and remained constant up to pH 11. The kinetics of Hg (II) adsorption on coirpith carbon followed Lagergren first order rate equation whereas adsorption equilibrium followed both Langmuir and Freundlich isotherm models. The adsorption capacity was found to be 154 mg/g at pH 5.0 for the particle size 250–500 mm.

Anirudhan et al. [15] produced a new cation exchanger from coconut coir pith, a coir industry based lignocellulosic residue, for removal of Hg (II) from simulated as well as from real industrial waste water in batch mode. The new adsorbent was synthesized by grafting poly(hydroxyethylmethacrylate) onto coconut coir pith, CP, using potassium peroxydisulphate as an initiator and in the presence of *N,N'*-methylenebisacrylamide as a cross-linking agent. The effect of pH, initial metal concentration, time, temperature, and ionic strength on adsorption were studied. Study expressed that increase in initial concentration accelerated the diffusion of Hg (II) ion from bulk solution to the surface of adsorbent due to the increase in driving force and as a result the amount of Hg (II) adsorbed at equilibrium increased from 4.97 mg/g (99.8%) to 32.90 mg/g (87.7%) when initial concentration increased from 10 to 75 mg/l. The maximum adsorption percentage of Hg (II) by the raw coir pith and the treated adsorbent was observed at the pH range 5.5–8.0 and significantly decreased by reducing the pH values to 2.0. The adsorption follows a first-order reversible kinetics. Lower ionic strength and high temperature were found to improve the uptake of Hg (II) from aqueous solution. The equilibrium isotherm data were fitted well by Freundlich adsorption isotherm model.

Bagasse pith, a common agricultural waste generated after the extraction of juice, was used by Krisnan and Anirudhan [16] to remove Hg (II) from the effluent coming out from chloro-alkali industries. After preliminary processing, the bagasse pith was carbonized at 200°C and steam activated. Depending on the activation condition, four different types of adsorbent *viz.*, steam activated carbon in presence of SO_2 and H_2S (SA–SO_2–H_2S–C), steam activated carbon in presence of SO_2 only (SA–SO_2–C), steam activated carbon in presence of H_2S only (SA–H_2S–C) and steam activated carbon only (SA–C) were prepared and used for removal of Hg (II) separately. They were compared with respect to their Hg (II) removal capability. pH, agitation time, ionic strength, particle size and temperature were considered as different parameters. The uptake of Hg (II) was maximum by SA–SO_2–H_2S–C followed by SA–SO_2–C, SA–H_2S–C and SA–C. The sulfurized activated carbon was effective in the pH range 4–9 whereas the sulfur free activated carbon was effective within a narrow range of pH 6–8. The amount of Hg (II) adsorbed increased with contact time and attained equilibrium at about 200 min for sulfurized activated carbons and 240 min for sulfur free activated carbons for an initial concentration of 100 mg/l Hg (II). It was also seen that metal ion adsorption was higher with smaller particle size. The metal sorption mechanism was controlled by pore diffusion process. Decrease in ionic strength and increase in temperature were found to favor the uptake of Hg (II). Study of adsorption isotherm reflected that the Hg (II) is strongly sorbed on the sorbent samples and there was no competition from the solvent for sorption sites. The adsorption equilibrium data were found to fit a Langmuir type isotherm model. The maximum adsorption capacity of Hg (II) (188.7 mg/g) was obtained with steam activation in presence of SO_2 and H_2S together (SA–SO_2–H_2S–C) and the adsorption capacity of the sulfur free activated carbon (Q_0) was reported to be 172.4 mg/g (SA–C). The complete removal of Hg (II) from 50 ml of synthetic and industrial wastewaters was achieved with 200 and 100 mg, 250 and 150 mg, 300 and 200 mg, and 400 and 300 mg of SA–SO_2–H_2S–C, SA–SO_2–C, SA–H_2S–C, and SA–C respectively.

11.3.3 Seeds and Nuts

A different type of agricultural residue like cotton seed was used by Srinivasan and Sathiya [17] for preparation of cotton seed carbon to be used for abatement of Hg (II) from waste water in batch mode. The collected cotton seed was carbonized using sulfuric acid as an activating agent at an elevated temperature. The influence of various factors such as agitation time, pH and carbon dosage upon sorption capacity were considered. Percentage removal of Hg (II) increased with contact time and equilibrium was attained within 3 hr. Maximum removal of Hg (II) took place at pH 4 to 5. It was evident that 99% removal could be achieved for an initial concentration of 10 mg/l of Hg (II) ions in 100 ml of solution with a carbon dosage of 100 mg. Langmuir was found to be the best for fitting the equilibrium data. Maximum 7.246 and 36.0 mg/g removal have been found to be possible from distilled water and tap water respectively. The sorption followed reversible first order kinetics where film diffusion was the rate determining step.

The possible use of tamarind nut, an agricultural waste, for the preparation of activated carbon to be used as Hg (II) removal agent was investigated by Ramadevi and Srinivasan [18] in batch mode contacting device. Bicarbonate treated tamarind nut carbon (BTNC) was prepared by carbonizing tamarind nut powder in presence of sulfuric acid and sodium bicarbonate. The adsorption of Hg (II) was examined under different parameters like time of agitation, pH, amount of carbon. The capacity of the prepared adsorbent was compared with commercial activated carbon and it was found that tamarind nut carbon is more effective in Hg (II) removal than commercial activated carbon. Within 3 hr the equilibrium was achieved for tamarind nut carbon with total removal of Hg (II) whereas for commercial activated carbon, equilibrium reached after 4 hr and only 92% Hg (II) had been removed. It was also found that 80 mg of tamarind nut carbon was sufficient for removal of 10 mg/l of Hg (II) from 100 ml solution while 150 mg of commercial activated carbon was required for removal of same amount of Hg (II). Tamarind nut carbon was effective in wide range of pH (4–10) in comparison with commercial activated carbon which was effective between narrow ranges (2–3). The adsorption capacities (Q_0) were reported as 60.24 and 62.5 mg/g for tamarind nut carbon and commercial activated carbon respectively. The adsorption of Hg (II) followed first order reversible kinetics and the rate determining step was film diffusion.

A summarized view of mercury removal by different agricultural wastes is presented in Table 2.

TABLE 2 Summarized view of removal of mercury by different agricultural wastes.

Adsorbent		Nature of Hg	Type of effluent	Reactor	Kinetic and Equilibrium model	Adsorption capacity	References
Precursor	Activating agent						
Rice husk	Alkali	Hg (II)	Simulated solution	Batch	Langmuir	33.1 mg/g	Krisnani et al. [7]
Rice husk	Sulfuric acid	Hg(II)	Simulated solution	Batch	PSOM, Langmuir	384.62 mg/g	El-Shafey [8]
Barley husk	Elemental sulfur	Hg^0	Hg^0/N_2 mixing gas	Fixed bed column	PSOM, Langmuir	310.9 μg/g	Hsi et al. [10]
Rice husk	Elemental sulfur	Hg^0	Hg^0/N_2 mixing gas	Fixed bed column	PSOM, Langmuir	344.3 μg/g	Hsi et al. [10]
Coconut husk	–	Hg (II)	Simulated aqueous solution	Batch	Freundlich	–	Hasany and Ahmad [11]
Ceiba pentandra hulls	Steam	Hg (II)	Waste water	Batch	PSOM, Freundlich	29.20 mg/g	Rao et al. [12]
Phaseolus aureus hulls	Steam	Hg (II)	Waste water	Batch	PSOM, Freundlich	23.37 mg/g	Rao et al. [12]
Cicer arietinum	Steam	Hg (II)	Waste water	Batch	PSOM, Freundlich	21.49 mg/g	Rao et al. [12]
Coir pith	Zinc chloride	Hg (II)	Simulated solution	–	–		Namasivayam and Sangeetha [13]
Coconut coir pith	Ammonium persulphate	Hg (II)	Waste water	Batch	Lagergren first order rate equation, Langmuir	154 mg/g	Namasivayam and Kadirvelu [14]
Coconut coir pith	Potassium peroxydisulphate and *N,N'*-methylenebisacrylamide	Hg (II)	Simulated and industrial waste water	Batch	First order reversible kinetics, Freundlich	32.90 mg/g	Anirudhan et al. [15]
Bagasse pith	Sulfur dioxide and Hydrogen sulphide	Hg (II)	Chloro-alkali industrial waste	Batch	Langmuir type isotherm	188.7 mg/g	Krisnan and Anirudhan [16]

TABLE 2 *(Continued)*

Adsorbent		Nature of Hg	Type of effluent	Reactor	Kinetic and Equilibrium model	Adsorption capacity	References
Precursor	Activating agent						
Cotton seed carbon	Sulfuric acid	Hg (II)	Waste water	Batch	First order reversible kinetics, Langmuir	36.0 mg/g	Srinivasan and Sathiya [17]
Tamarind nut carbon	Sulfuric acid and Sodium bicarbonate	Hg (II)	Simulated solution	Batch	First order reversible kinetics	60.24 mg/g	Ramadevi and Srinivasan [18]

11.4 ANIMAL WASTES

Though reclamation of animal wastes for preparation of adsorbents is not very common like agricultural wastes, few investigations have been made to explore the possibility of uses of different animal wastes like hooves, carapace, bones, and so on depending on the local availability of these materials.

11.4.1 Hooves, Carapace, and Bones

Touaibia and Benayada [19] made an attempt to explore the possibility of uses of algerian sheep hooves for the preparation of Keratin powder, an adsorbent, to be used for removal of mercury (II) from aqueous solution. After preliminary processing, algerian sheep hooves were steamed and dried and after then these were ground into required particle size (<63 μm). All the experiments were performed in batch mode under varying pH, sorption time and initial concentration of metal. Initial pH (2–10) had no effect on adsorption of mercury on sheep hooves powder. Effect of initial concentration was examined in the concentration rage of 1–100 mg/l of Hg (II). Increase in initial concentration facilitated the uptake of Hg (II). Kinetics of adsorption of Hg (II) on the adsorbent followed pseudo-second-order kinetic model and equilibrium data fitted to Freundlich adsorption isotherm model.

The potential of carapace from the edible crab was assessed for the adsorption of Hg (II) from aqueous solutions by Rae et al. [20]. Influence of pH, contact time, initial concentration of metal, particle size and effect of co-ion (Cu(II)) on Hg (II) sorption was examined in batch study. The Hg uptake increased from 0.01 to 13.0 mg/g as initial concentration increased from 0.5 to 1000 mg/l. Within first 40 min, uptake of metal increased to 7.8 mg/g sharply whereas a slow uptake was observed afterward reaching 8.7 mg/g in 240 min. The initial Hg uptake was rapid and proceeds *via* physicochemical interactions at external surfaces depending on the particle size. The Hg (II) uptake increased from 0.01 to 13.0 mg/g as initial concentration increased from 0.5 to 1,000 mg/l. When the particle size was reduced from >2.5 to <0.15 mm, Hg (II) uptake increased from 1.41 to 8.43 mg/g. The quick availability and durability in granular forms made the crab carapace a strong contender as low cost alternatives for adsorption of Hg (II).

Camel bone, another type of animal waste, was used by Hassan et al. [21] as a precursor for preparation of charcoal to remove mercury from waste water. The bone charcoal (BC) residue was the resultant of a pyrolysis process at an elevated temperature (800°C). Adsorption of Hg (II) using BC was carried out in a batch mode contacting device and pH, contact time and amount of camel BC was taken as parameters. Utmost 10 mg/l of Hg (II) can be removed by at least 0.03 g of adsorbent. Equilibrium was achieved within 30 min of contact with stirring. The effect of pH was examined in the pH range 2 – 7. The maximum adsorption was observed at pH 2.0. Equilibrium studies showed that adsorption of mercury (II) on camel BC followed Langmuir adsorption isotherm model. The adsorption capacity (Q_0) was found to be 28.24 mg of Hg (II)/g of the adsorbent. The optimum removal conditions were reported as pH 2, contact time 30 min and temperature 25°C. Adsorption of Hg (II) from waste water collected from research laboratories (0.1–5 μg/ml) and from separator tanks of crude petroleum oil (0.05–2 μg/ml) resulted in 95.8–98.5% removal of the original mercury concentrations. Calcium phosphate, a major component of camel BC, acted not only as a source of adsorption centers but also enables ion-exchange process.

The mechanism of Hg (II) uptake, suggested by Hassan et al. [21] is as follows:

$$\equiv POH + M^{2+} = \equiv POM^{+} + H^{+}$$

$$\equiv PO^{-} + M^{2+} = \equiv POM^{+}$$

$$\equiv CaOH + M^{2+} = \equiv CaOM^{+} + H^{+}$$

Where M denotes Hg.

A summarized view of mercury removal by different animal wastes is presented in Table 3.

TABLE 3 Summarized view of removal of mercury by wastes from animal origin.

Adsorbent		Nature of Hg	Type of effluent	Reactor	Kinetic and equilibrium model	Adsorption capacities	References
Precursor	**Activating agent**						
Keratin powder from algerian sheep hooves	–	Hg (II)	Simulated solution	Batch	PSOM, Freundlich	68 mg/g	Touaibia and Benayada [19]
Crab carapace	–	Hg (II) associated with Cu (II)	Simulated solution	Batch	–	13.0 mg/g	Rae et al. [20]
Camel bone charcoal	Pyrolysis	Hg (II)	Waste water	Batch	Langmuir	28.24 mg/g	[21]

11.5 INDUSTRIAL WASTES

Though in true sense most of the agricultural wastes as categorized earlier may be obtained from different agro-industry, they are actually generated as the residue left over after physical separation. Whereas the category "Industrial wastes" include the wastes generated after chemical processes of the substrates. Thereby, rice husk ash (RHA), though it is biomass in origin, has been included in the category of industrial wastes. Again, combustion of coal generates tons of fly ash in thermal power plants. As it consists of several oxides like silica, alumina, and so on having the capability of adsorption of several pollutants, this can be used as an efficient adsorbent to treat waste water. Thus, studies utilizing fly ash as an adsorbent for removal of mercury either directly or indirectly are reviewed in this section. Furthermore, some investigations where wastes generated from tyre industry and sago and palm oil industry have been reported to be used as mercury removal agent are also reviewed in this chapter.

11.5.1 Rice Husk Ash

Another attempt to use RHA for effective removal of Hg (II) was tested by Tiwari et al. [22]. RHA with diameter of 130–600 μm was used in batch experiment as well as in continuous flow experiment. Effect of initial concentration, contact time, particle size and pH on adsorption of Hg (II) was studied. Adsorbed concentration of Hg (II) on RHA increased from 2.3 to 6.8 mg/g when initial concentration was raised from 4.0 to 364.0 mg/l. Adsorption decreased considerably (98–74%) when particle size was increased. Maximum adsorption occurred at pH 6. Breakthrough studies were carried out by passing Hg (II) solution having initial concentration of Hg (II) 100 mg/l at pH 5.5, through the column packed with 50 g of RHA at a flow rate of 30 ml/min. It was observed that 8 bed volumes could be passed through the column without any Hg (II) being detected in the effluent. Lower flow rate and larger bed height favored the uptake of Hg (II) from solution as evident from column study.

Zhao et al. [23] showed the potential of iodine modified RHA as an adsorbent to remove elemental mercury. Combustion of rice husk was performed at 600°C for 4 hr. Sorption experiments of vapor phase elemental mercury (Hg^0) were carried out in a laboratory scale fixed-bed reactor. The effects of I_2 concentration (5–10%), supports, pore size distribution, impregnation manner of I_2 and temperature (80–140°C) on the Hg^0 capture efficiency were investigated. Study revealed that at 10% loading of I_2, 80% of Hg^0 was removed by adsorption. It was found that the surface area of support, pore size distribution and impregnation manner of I_2 had no effect on adsorption of Hg^0 whereas the I_2 content was the most important factor influencing the Hg^0 adsorption capacity. The adsorption of Hg^0 increased with increase in temperature and reached to almost 100% at above 140°C temperature. The experiment under a wide range of temperature implied that chemisorptions played an important role in Hg^0 removal. The reaction mechanism proposed by Zhao et al. [23] for Hg^0 capture by sorbents is shown below:

$$M^0 + I_2 \Leftrightarrow M....I_2*$$

$$M....I_2* + S \Leftrightarrow MI_2(ads)$$

$$MI_2(ads) + Ca - Site \Leftrightarrow MI_2Ca - Site$$

Here *M* stands for Hg and *S* stands for any gaseous molecule or solid surface in the reaction system.

London dispersion forces between Hg^0 and iodine acted in the first step of the process led to the formation of transient species (e.g., $Hg \cdots I_2^*$). The collision of $Hg...I_2^*$ with '*S*' might be taken place in second step where '*S*' could be the solid surface or any gaseous molecule in the reaction system. Thereby, a stable product HgI_2 was formed. Experiments revealed that the physical properties of sorbents did not affect Hg^0 adsorptive capacity. The reaction of sorbed HgI_2 to the alkaline site of Ca-based sorbents (Ca-Site) occurred in the third step of the process and thus binary iodide was formed.

The capability of RHA, as an adsorbent for abatement of mercury (II), was also explored by El-Said et al [24]. Like previous one, here also adsorption of mercury was carried out in a batch contactor. Author considered a competitive adsorption of Cd (II) and Hg (II) on RHA from a single and binary system. Three input parameters *viz.*, pH (2–8), adsorbent dosage (1–12 g/l) and initial metal ion concentration were varied to assess their effect on Hg (II) removal. Adsorption of metal increased gradually with pH upto pH 4 and a drastic increase in adsorption continued up to pH 6. After that the adsorption became constant. Removal of metal ions increased with an increase in the adsorbent dosage from 1 to 10 g/l. The removal remained unchanged above 10 g/l of RHA. An increase in the initial concentration of the metal ions enhanced the adsorption of Hg (II) ions onto RHA. The rate of metal ion removal was found to be very rapid during the initial 15 min after that the rate of metal ion removal decreased. No considerable change in metal ion removal was observed after about 120 min. The adsorption equilibrium data best fitted to the Freunlich adsorption isotherm model. The highest adsorption capacity as reported by El-Said et al [24] was 3.401 mg/g of adsorbent when metal concentration was 100 mg/l.

Polyaniline nanocomposite coated RHA, an unusual adsorbent, was used for removal of Hg (II) by Ghorbani et al. [25]. RHA was obtained by burning of rice husk. It was then treated with KIO_3 and sulfuric acid. Completely mixed batch reactor (CMBR) was employed to remove Hg (II) from waste water using three adsorbents *viz.*, polyaniline, RHA and polyaniline/RHA nanocomposite separately. Experiments were carried out to evaluate the influence of pH, contact time, adsorbent dosage and rotating speed. Initial pH was varied from 2 to 12. Removal of mercury increases with increase in solution pH and a maximum value was reached at an around pH 9. The removal efficiency was higher for polyaniline naocomposite coated rice husk than the other two adsorbents. The removal of mercury ions increased with every increment in the rotating speed (varied from 100–800 rpm) up to 400 rpm, afterward percentage removal decreased. Thus, 400 rpm was preferred as an optimized rotating speed. Amount of adsorbent was also varied from 100–1000 mg at 100 mg/l of Hg (II) concentration. The maximum removal was obtained at an optimum amount of adsorbent after which increase in adsorbent did not influence the result. Study revealed that adsorption of mercury was fast and the equilibrium was achieved after 20 min of contact time using all three adsorbents. The results revealed that the PAn/RHA nanocomposite has an ample potential for the removal of Hg (II) from aqueous solution. Optimum conditions for mercury removal were found to be at pH 9, adsorbent dosage of 10 g/l, contact time 20 min and rotating speed 400 rpm.

11.5.2 Fly Ash

The major solid waste of thermal power plants based on coal burning is fly ash. The main uses of fly ash include construction of roads, bricks, cement and so on. The high percentage of silica and alumina in fly ash makes it an effective adsorbent for bulk use.

Adsorption characteristics of unburned carbons from fly ash in terms of elemental mercury removal were examined by Li et al. [26]. Unburned carbons from six different fly ash sources were studied. Both batch and continuous studies were made in the present investigation. The batch adsorption tests indicated that the unburned carbons from fly ash had significant adsorption capacities for elemental mercury. For unburned carbons, the mercury adsorption capacity could be as high as 60 mg/g. While, heat treatment had a significant impact on the adsorption capacity of the unburned carbon, adsorption temperature had adverse effect on adsorption capacity. It was seen that adsorption capacity at 40°C was lower than that at 20°C. Compared to the capacity obtained in batch tests, the capacity in the column tests was found to be much lower, less than 10% of that from the batch tests. The mechanism of adsorption of mercury on carbon was explained as the physical and chemical interactions taking place between the carbon surface and mercury.

Fly ash had also been used by Rio and Delebarre [27] as an adsorbent to remove mercury from aqueous solution using fluidized bed plants. Depending on the composition of the fly ash, it was categorized as silico-aluminous ash, consisting of 65% silica and alumina and the sulfo-calcic ash, containing high content of sulfur trioxide and calcium oxide. Adsorption experiment revealed that equilibrium reached within 72 hr for both type of fly ashes but for the sulfo-calcic fly ash, the rate of adsorption was faster and steady state removal was higher. Experiments revealed that adsorption was favorable at low initial concentration of Hg (II). When initial concentration was much higher, adsorption capacity remained unaltered. The maximum adsorption capacities at equilibrium were equal to 3.2 mg/g for silico-aluminous ashes and 4.9 mg/g for sulfo-calcic ashes, corresponding to removal efficiencies of 53 and 81%, respectively. Furthermore, leaching of mercury from silico-aluminus fly ash was more than the sulfo-calcic one. This proves that adsorption on sulfo-calcic fly ash was more stable than silico-aluminous fly ash.

Fly ash is a very popular adsorbent to remove not only mercury but all heavy metals, dyes and different pollutants. Somerset et al. [28] synthesized zeolites from fly ash and co-disposal filtrate to be used as adsorbent for removal of mercury. Co-disposal filtrates were obtained by a co-disposal reaction. Raw fly ash and co-disposal filtrates were used as a precursor for alkaline hydrothermal zeolite synthesis. The sample was incubated with sodium hydroxide (NaOH) in a 1:1.2 ratio, at 600°C for approximately 1–2 hr. The fused product was then mixed thoroughly with distilled water and the slurry was subjected to aging for 8 hr. After aging, the slurry was subjected to crystallization at 100°C for 24 hr. In subsequent step, the solid product was recovered by filtration and washed thoroughly with deionizer water until the filtrate had a pH of 10–11. The recovered product was then dried at a temperature of 70°C. Better removal of mercury was obtained using zeolites prepared from co-disposal filtrates. Synthesized zeolite material was effective in reducing the Hg (II) concentrations from 0.47 to 0.17 mg/kg (30%).

11.5.3 Tire Granules

Tire industry generates lots of tire granules as wastes. This waste material can effectively be used for the purpose to remove pollutants.

Performance of non-porous tire granules as an adsorbent was investigated by Danwanichakul and Danwanichakul [29]. The material used in this study was size-reduced residual rubber obtained from tire wastes. Batch mode was selected for experimentation. In case of adsorption of Hg (II) on the tire granules, film diffusion was considered as rate limiting step. It was found that adsorption rate increased with rubber loading and the adsorption rate was greater for smaller size particles. A model was proposed to estimate the mass diffusivity of Hg (II) in aqueous solution to calculate the mass transfer coefficient which is necessary for designing of the adsorption column where the effects of the convective transfer and diffusion are present. The adsorption reaction was best described by the pseudo-second order reaction. To get the concentration profile with time, Danwanichakul and Danwanichakul [29] proposed the following kinetic equation assuming the film diffusion as the rate controlling step for adsorption of Hg (II) on waste tyre particles.

$$\ln\frac{C}{C_0} = -\frac{12mD_{ab}}{V\rho_p D_p^2}t$$

Where,

C_0 = Initial concentration of Hg (II) solution, C = Concentration of Hg (II) solution at any time, m = Total mass of waste tire rubber, D_{ab} Mass diffusivity of Hg (II) in water, V = Solution volume, ρ = Density, t = Time, D_p = Particle diameter.

11.5.4 Sago and Palm Oil Industry Wastes

Activated carbon made from sago industry wastes either by chemical or physical activation method as a promising adsorbent to remove Hg (II) was described by Kadirvelu et al. [30]. The material was charred using H_2SO_4 and $(NH_4)_2S_2O_8$ at room temperature for 0.5 hr. In batch mode studies, the adsorption was dependent on particle size, solution pH, initial Hg (II) concentration and carbon dosage. Percentage removal was found to be decreased with increasing particle size. When carbon dosage was increased, Hg (II) removal was also increased. Equilibrium study showed that adsorption of Hg (II) on activated carbon made from sago industry generated wastes followed Langmuir adsorption isotherm model. The adsorption capacity of the activated carbon made from sago industry wastes was reported as 55.6 mg/g. The removal efficiencies of metal ion were affected by the initial metal ion concentration. Removal efficiency decreased with the increase in metal ion concentration at constant pH. Based on the results obtained, author concluded that sago carbon could be an effective adsorbent for the removal of Hg (II) from aqueous solutions.

An attempt to prepare activated carbon from palm oil empty fruit bunch was made by Wahi et al. [31]. Dried palm oil empty fruit bunch was carbonized at 400°C for 30 min. Then the material was soaked in NaOH at 60°C. The material was carbonized for the second time at 700°C for 1 hr and treated with 5M HCl. The Hg (II) adsorption capacity of this activated carbon was found to be 52.67 mg/g. The adsorption of

Hg (II) on activated carbon was dependent on the adsorbent dosage and initial metal concentration. When the initial metal concentration was varied between 5–20 mg/l, the removal reached 100%. Equilibrium studies showed that adsorption of Hg (II) was best explained by Fruendlich model. The study revealed that the activated carbon prepared from palm oil empty fruit bunch through chemical activation using NaOH as the activating agent could be used as an effective adsorbent for the removal of Hg (II) from waste water.

A summarized view of mercury removal by different industrial wastes is presented in Table 4.

TABLE 4 Summarized view of removal of mercury by different industrial wastes.

Adsorbent		Nature of Hg	Type of effluent	Reactor	Kinetic and equilibrium model	Adsorption capacities	References
Precursor	Activating agent						
Rice husk ash	–	Hg (II)	Simulated solution	Batch and continuous flow	Langmuir	6.8 mg/g	Tiwari et al. [22]
Rice husk ash	Iodine	Hg^0	Vapor phase	Fixed bed reactor	–	–	Zhao et al. [23]
Rice husk ash	–	Hg (II) in presence of Cd (II)	Simulated solution	Batch	Freundlich	3.401 mg/g	El-Said et al. [24]
Rice husk ash	Polyani-line	Hg (II)	Simulated solution	Completely mixed batch reactor		–	Ghorbani et al. [25]
Unburned carbon from fly ash	–	Hg^0	Simulated gaseous effluent	Batch and continuous mode	–	60 mg/g	Li et al. [26]
Fly ash	–	Hg (II)	Simulated solution	Fluidized bed	–	4.9 mg/g	Rio and Delebarre [27]
Zeolites from fly ash	Sodium hydroxide	Hg (II)	Waste water	Batch	–	0.17 mg/kg	Somerset et al. [28]
Non-porous tire granules	–	Hg (II)	Simulated solution	Batch	PSOM	–	Dan-wanichakul and Dan-wanichakul [29]
Sago industry wastes	Sulfuric acid and am-monium persulfate	Hg (II)	Simulated solution	Batch	Langmuir	55.6 mg/g	Kadirvelu et al. [30]
Activated carbon from palm oil empty fruit bunch	Sodium hydroxide	Hg (II)	Simulated solution	Batch	Freundlich	52.67 mg/g	Wahi et al. [31]

11.6 MISCELLANEOUS WASTES

The types of waste *viz.*, wood, leaves and so on which do not belong to the previous categories as such but have been used for making adsorbent for removal of mercury have been listed below.

11.6.1 Various Tree-parts

Though commercial activated carbon is a widely accepted adsorbent for mercury removal, high initial, and regeneration costs restrict its use in wide scale. Nowadays researches are concentrating on finding new precursors to manufacture activated carbon and thereby minimizing the cost of the process. Eucalyptus wood is one of such precursor which was used to prepare activated carbon by Silva et al. [32]. The material was carbonized in absence of air at 773K for 2 hr. The carbonized material (CEW) was then activated with water vapor. Surface modification was done by sulfurization using sulfuric acid and carbon disulphide separately. When sulfuric acid was used as a sulfurizing agent, the adsorbent was denoted by $AC\_H_2SO_4$ and in second case activated carbon was designated by AC_CS_2 when carbon disulphide was used. Effect of initial concentration of metal and adsorbent dosage was studied extensively. Equilibrium adsorption data were fitted satisfactorily to Freundlich adsorption isotherm model which suggest that the surface modification with sulfuric acid created important changes on the structure of the adsorbent. The sulfurization treatment with carbon disulphide also led to important changes in both, textural properties and surface chemistry of the solid. In this case, the decrease in the surface area and pore volume was balanced by a favorable surface chemistry for mercury adsorption which made this adsorbent better than the untreated solid.

Similar effort was made by Zabihi et al. [33] as they used walnut shell as a precursor for the preparation of activated carbon. The powered sample was treated with $ZnCl_2$ as an activating agent in weight ratio of shell to $ZnCl_2$ as 1:0.5 (Carbon A) and 1:1 (Carbon B). Adsorption of Hg (II) from aqueous solutions was examined under different experimental conditions by varying contact time, metal ion concentration, pH and solution temperature. The adsorption capacity increased with the increase in initial Hg (II) concentration from 9.7 to 107 mg/l. It was observed that removal efficiency of Hg (II) reduced with increasing pH. Metal uptake by the adsorbent increased with increase in temperature which suggests that the reaction is endothermic in nature. The adsorption kinetics was well described by pseudo-second-order kinetic model and equilibrium data were fitted to Langmuir isotherm model. The adsorption capacity of new adsorbents were reported as 151.5 mg/g (Carbon A) and 100.9 mg/g (Carbon B) at pH = 5.0.

Bamboo can be an important precursor for activated carbon. Tan et al. [34] used both bamboo charcoal and bamboo charcoal modified with H_2O_2 for removal of elemental mercury. The pretreated BC materials were impregnated with oxydol solution. Depending on the concentration of oxydol solution the prepared material was designated as BC1 and BC2. Influence of temperature and presence of O_2 were studied extensively. The efficiency of the adsorbent was examined at two temperatures like 20°C and 90°C. The adsorption quantity of BC1 and BC2 materials were 1470.5 and 1347.9 μg/g at 20°C and 103.5 and 219.7 μg/g at 90°C respectively. Reports revealed that mercury uptake by the prepared adsorbent can be enhanced by the presence of oxygen. Gaseous mercury can be oxidized by oxygen in flue gas and oxidized mercury can be easily adsorbed than its unoxidized form. The BC materials after modification with H_2O_2 acquired an excellent adsorption capacity for elemental mercury.

Pulido et al. [35] have carbonized sugi (*Cryptomeria japonica* D. Don) wood powder in order to remove mercury from aqueous solution. Wood powder from sugi were carbonized in a nitrogen atmosphere at varying furnace temperatures of 200, 600, and 1000°C for 1 hr and termed as C200, C600, and C1000 respectively. Mercury was removed to a greater extent by C1000 and C600 than by C200 or wood powder in its native form. The ability of non-carbonized wood powder to remove the metals could be accredited to the original nature of the raw wood and an increased specific surface area due to swelling in water. Carbonized wood powder removed almost all the mercury after 1 hr of operation, whereas it took about 24 hr for commercial activated carbon to reach the same level of removal. Wood powder carbonized at 1000°C performed best among the wood charcoals.

In 2011, Wahby et al. [36] used olive stones to prepare activated carbon. They treated crushed olive stones with sulfuric acid and then it was carbonized under nitrogen gas flow. Samples were activated in two ways. Some of the samples were subjected to pre-oxidation prior to activation at high temperature under CO_2 flow and rest of the sample was only activated in high temperature under CO_2 flow without any oxidation. It was found that Hg (II) removal capacity of activated carbon prepared without the pre-oxidation process was much less than the pre-oxidized carbon. Seventy two percentage removals were achieved with pre-oxidized carbons when the initial concentration of mercury was 20 mg/l at pH 2. The Hg (II) adsorption kinetics on pre-oxidized carbon followed first order rate expression given by Lagergren.

Date palm leaflet was used by El-Safey and Al-Haddabi [37] as a precursor for activated carbon. The sample was carbonized using sulfuric acid as a chemical activating agent. Adsorption was found to depend strongly on pH. Sorption of Hg (II) was accompanied by a reduction in the final pH representing release of protons in solution. This implied an ion exchange mechanism to be taking place in the adsorption process. The equilibrium for Hg (II) adsorption reached within 120 hr. Investigation revealed that the kinetics of adsorption followed pseudo second order model suggesting chemisorptions as the rate determining step. Equilibrium data fitted well to Langmuir equation. The maximum removal (274 mg/g) of Hg (II) was obtained at 45°C.

Guava bark can be an effective adsorbent for Hg (II) adsorption as proposed by Lohani et al. [38]. Collected guava barks were dried and crushed to desired size. The adsorption of mercury was found to be dependent on the initial concentration of Hg (II). With an increase in initial concentration, adsorption increased. When pH of the solution was changed from acidic to alkaline, adsorption capacity increased from 2.292 to 3.364 mg/g and maximum adsorption took place at pH 9.0. Temperature had a direct effect on amount of adsorption. The negative values of ΔG^0 reflected the nature of process is spontaneous and the positive ΔH^0 specified the endothermic adsorption process. More than 60% adsorption took place in the first 60 min of operation. Adsorption kinetics followed pseudo second order kinetic model. The maximum adsorption capacity was found to be 3.364 mg/g when initial concentration was 50 mg/g and temperature was 323K. Adsorption equilibrium data were fitted to Freundlich isotherm at various temperatures.

A summarized view of mercury removal by miscellaneous wastes is presented in Table 5.

TABLE 5 Summarized view of removal of mercury by miscellaneous wastes.

Adsorbent		Nature of Hg	Type of effluent	Reactor	Kinetic and equilibrium model	Adsorption capacities	References
Precursor	Activating agent						
Eucalyptus wood carbon	Sulfuric acid and carbon disulphide	Hg (II)	Simulated solution	Batch	Freundlich	–	Silva et al. [32]
Wall nut shell carbon	Zinc chloride	Hg (II)	Simulated solution	Batch	PSOM, Langmuir	151.5 mg/g	Zabihi et al. [33]
Bamboo charcoal	Hydrogen peroxide	Hg^0	Simulated flue gas	–	–	1470.5 μg/g	Tan et al. [34]
Carbonized wood powder from suji	–	Hg (II) with other heavy metals	Simulated solution	–	–	–	Pulido et al. [35]
Olive stones	Sulfuric acid	Hg (II)	Simulated solution	–	Lagergren first order rate expression	–	Wahby et al. [36]
Date palm leaflet	Sulfuric acid	Hg (II)	Simulated solution	–	PSOM, Langmuir	274 mg/g	El-Safey and Al-Haddabi [37]
Guava bark	–	Hg (II)	Simulated solution	Batch	PSOM, Freundlich	3.364 mg/g	Lohani et al. [38]

11.6.2 Recovery and Recycle of Mercury and Reuse of Adsorbent

Any environmental pollution removal process cannot be completed unless the removed pollutants are either destroyed or recovered and recycled in the process itself or reused in any other process. If this does not occur, the process cannot be categorized as pollutant abatement process but can be termed as a transfer process where pollutant is transferred from one phase to another phase retaining the same problem of pollution. Therefore, recovery of pollutant vis-à-vis regeneration of matrix is mandatory in adsorption from environmental pollution abatement point of view. In the context of the present paper, recovery of adsorbed mercury vis-à-vis regeneration of matrix is very much essential. Regeneration of the adsorbent minimizes the operating cost and recovered metal can be recycled in industry. Otherwise the adsorbent with bound metal would add to the environmental pollution due to the disposal problem. Unfortunately, only very few chapter discussed in the foregoing sections deal with the recovery study and they are discussed in the following section.

A study regarding removal and recovery of mercury was performed by Namasivayam and Kadirvelu in 1999 using coconut coir pith carbon as an adsorbent. Desorption of Hg (II) ion from prepared matrix was examined using different strengths of HCl and KI. The maximum percent recovery of Hg (II) was reported to be 63% with 0.5 M HCl and 84.0% with 2.0% KI solution. Higher desorption of Hg (II) by I^- might be due to the formation of relatively more stable iodide complexes with Hg (II) compared to that of chloride complexes.

A new ion exchanger made from coconut coir pith was employed by Anirudhan et al. [15]. The reusability of adsorbent after recovery of adsorbed mercury (II) was examined in the present study. Since ion exchange is generally reversible process, the Hg (II) desorption performance was examined using HCl solution. Desorption increased with increasing HCl concentration. The formation of chloro-complexes with Hg (II) is the main reason for desorption of Hg (II). Maximum 98.3% desorption of mercury (II) was obtained using 0.2 M HCl. After four consecutive cycles the adsorption capacity of the adsorbent reduced from 99.3 to 90.4% while the recovery of Hg (II) decreased from 98.7 to 91.4% in the fourth cycle.

Krisnan and Anirudhan [16] used sulfurized bagasse pith to remove Hg (II) from chloro-alkali industries. To examine the reusability of the prepared adsorbent, desorption studies of Hg (II) was carried out with 0.2 M HCl and more than 95% of Hg (II) was recovered in this process. After three successive cycles almost 91–97% of the initial desorption capacity was achieved by the investigators.

El-Said et al [24] studied the regeneration of RHA to recover bound Hg (II) from it. The study was carried out in batch mode with a number of solvents. When acetic acid was used as solvent, only 24.58% mercury (II) was recovered whereas with mineral acids like HCl, H_2SO_4, and HNO_3, recovery efficiency increased to approximately 31.5%. The desorption efficiency was not very significant in the present case. Thus, it may be said that the chemisorptive adsorption of metal ions onto RHA hindered the desorption of metal ions from the spent RHA.

11.7 CONCLUSION

In the present chapter an attempt has been made to review the investigations based on removal of mercury using adsorbent prepared from different types of waste materials. The specialty of such investigations lies in the fact that they are not only able to eradicate mercury from effluent but also reduce solid waste by reclaiming waste materials for preparation of low cost adsorbent and thereby lessening environmental pollution as a whole. The waste materials have been categorized as agricultural, animal, industrial and miscellaneous type depending on their origin. A thorough review has been made to compile the data obtained by different scientists worked in this line. It is to be noted that the adsorption capacity mentioned in the present chapter depends on experimental condition. However, some shortcomings are observed during the review process. They may be summarily presented as follows:

(i) To prepare adsorbent at low cost with high efficiency, the operating condition for preparation of adsorbent should be optimized in terms of its process variables.
(ii) Cost analysis is required to examine the actual expense of the process.

(iii) Though a number of works have been done to eradicate mercury using adsorbent prepared from agricultural waste, the utilization of industrial waste for the same is rather scarce in quantity. Thus, studies should be encouraged in this line.
(iv) Surface chemistry of adsorbent should be studied in great detail to predict the mechanism of interaction between adsorbate and adsorbent and thereby, selection of adsorbent will be easier for a specific adsorbate.
(v) Generally real waste water is accompanied not only with other metals but also with organic contaminants like dyes, phenols and so on. Such contaminants affect the rate and extent of adsorption of key pollutant over which the investigation was initially made. Thus, investigation should be carried out with simulated solution consisting of such pollutants and real industrial effluent as well.
(vi) Most of the studies reported here have been carried out in laboratory scale contactor operating in batch mode. But to handle huge amount of industrial effluent, continuous contactor is the only solution. Thus, experiments should be carried out in continuous contactor also. Pilot plant study using real industrial waste water is very much necessary to assess the feasibility of the prepared adsorbent in industrial scale.
(vii) Efficiency of adsorption process depends not only on the properties of adsorbent and adsorbate, but also on the operating condition of the system. Thus, judicial selection of various input parameters like initial concentration of adsorbate, weight of adsorbent, size of adsorbent, pH, temperature and so on, is necessary.
(viii) Studies on recovery of adsorbed metal and regeneration of adsorbent should be done in great detail to assess the effectiveness of the process in true sense.

KEYWORDS

- **Agricultural Waste**
- **Bone Charcoal**
- **Coconut Husk**
- **Equilibrium**
- **Langmuir Equation**
- **Mercury Pollution**

ACKNOWLEDGMENT

We acknowledge CSIR (Council of Scientific and Industrial Research, INDIA) for funding the fellowship of Ms. Aparupa Bhattacharyya.

REFERENCES

1. Dutta, S., Bhattacharyya, A., De, P., Ray, P., and Basu, S. Removal of mercury from its aqueous solution using charcoal-immobilized papain (CIP). *Journal of Hazardous Materials*, **172**, 888–896 (2009).

2. Agarwal, H., Sharma, D., Sindhu, S. K., Tyagi, S. and Ikram, S. Removal of mercury from wastewater use of green adsorbents—a review. *EJEAFChe*, **9**(9), 1551–1558 (2010).
3. Malachowski, L., Stair, J. L., and Holcombe, J. A. Immobilized peptides/amino acids on solid supports for metal remediation. *Pure and Applied Chemistry*, **76**(4), 777–787 (2004).
4. Bhatnagar, A. and Sillanpää, M. Utilization of agro-industrial and municipal waste materials as potential 2 adsorbents for water treatment—a review. *Chemical Engineering Journal*, doi:10.1016/j.cej.2010.01.007 (2008).
5. Dutta, S., Bhattacharyya, A., Ganguly, A., Gupta, S., and Basu, S. Application of Response Surface Methodology for preparation of low-cost adsorbent from citrus fruit peel and for removal of methylene Blue. *Desalination*, **275**, 26–36 (2011).
6. Bailey, S. E., Olin, T. J., Markbricka, R., and Adrian, D. D. A review of potentially low-cost sorbents for heavy metals. *Water Research*, **33**(11), 2469–2479 (1999).
7. Krishnani, K. K., Meng, X., Christodoulatos, C., and Boddu, V. M. Biosorption mechanism of nine different heavy metals onto biomatrix from rice husk. *Journal of Hazardous Materials*, **153**, 1222–1234 (2008).
8. El-Shafey, E.I. Removal of Zn (II) and Hg (II) from aqueous solution on a carbonaceous sorbent chemically prepared from rice husk. *Journal of Hazardous Materials*, **175**, 319–327 (2010).
9. Huang, L., Xiao, C., and Chen, B. A novel starch-based adsorbent for removing toxic Hg (II) and Pb (II) ions from aqueous solution. *Journal of Hazardous Materials*, **192**, 832–836 (2011).
10. Hsi, H. C., Tsai, C. Y., Kuo, T. H., and Chiang, C. S. Development of low-concentration mercury adsorbents from biohydrogen-generation agricultural residues using sulfur impregnation. *Bioresource Technology*, **102**, 7470–7477 (2011).
11. Hasany, S. M. and Ahmad, R. The potential of cost-effective coconut husk for the removal of toxic metal ions for environmental protection. *Journal of Environmental Management*, **81**, 286–295 (2006).
12. Rao, M. M., Reddy, D. H. K. K., Venkateswarlu, P., and Seshaiah, K. Removal of mercury from aqueous solutions using activated carbon prepared from agricultural by-product/waste. *Journal of Environmental Management*, **90**, 634–643 (2009).
13. Namasivayam, C. and Sangeetha, D. Recycling of agricultural solid waste, coir pith: Removal of anions, heavy metals, organics and dyes from water by adsorption onto $ZnCl_2$ activated coir pith carbon. *Journal of Hazardous Materials*, **B135**, 449–452 (2006).
14. Namasivayam, C. and Kadirvelu, K. Uptake of mercury (II) from wastewater by activated carbon from an unwanted agricultural solid by-product: Coirpith. *Carbon*, **37**, 79–84 (1999).
15. Anirudhan, T. S., Divya, L., and Ramachandran, M. Mercury (II) removal from aqueous solutions and wastewaters using a novel cation exchanger derived from coconut coir pith and its recovery. *Journal of Hazardous Materials*, **157**, 620–627 (2008).
16. Krishnan, K. A. and Anirudhan, T. S. Removal of mercury (II) from aqueous solutions and chlor-alkali industry effluent by steam activated and sulfurized activated carbons prepared from bagasse pith: Kinetics and equilibrium studies. *Journal of Hazardous Materials*, **B92**, 161–183 (2002).
17. Srinivasan, K. and Sathiya, E. Bimetal Adsorption by cottonseed carbon: equlibrium and kinetic studies. *E-Journal of Chemistry*, **6**(4), 1167–1175 (2009).
18. Ramadevi, A. and Srinivasan, K. Agricultural solid waste for the removal of inorganics: Adsorption of mercury (II) from aqueous solution by tamarind nut carbon. *Indian Journal of Chemical Technology*, **12**, 407–412 (2005).
19. Touaibia, D. and Benayada, B. Removal of mercury (II) from aqueous solution by adsorption on keratin powder prepared from Algerian sheep hooves. *Desalination*, **186**, 75–80 (2006).
20. Rae, I. B., Gibb, S. W., and Lu, S. Biosorption of Hg from aqueous solutions by crab carapace. *Journal of Hazardous Materials*, **164**, 1601–1604 (2009).
21. Hassan, S. S. M., Awwad, N. S., and Aboterika, A. H. A. Removal of mercury (II) from wastewater using camel bone charcoal. *Journal of Hazardous Materials*, **154**, 992–997 (2008).
22. Tiwari, D. P., Singh, D. K., and Saksena, D. N. Hg (II) adsorption from aqueous solutions using rice-husk ash. *Journal of Environmental Engineering*, 479 (1995).

23. Zhao, P., Guo, X., and Zheng, C. Removal of elemental mercury by iodine-modified rice husk ash sorbents. *Journal of Environmental Sciences*, **22**(10), 1629–1636 (2010).
24. El-Said, A. G., Badawy, N. A., and Garamon, S. E. Adsorption of cadmium (II) and mercury (II) onto natural adsorbent rice husk ash (RHA) from aqueous solutions: Study in single and binary system. *Journal of American Science*, **6**(12), 400–409 (2010).
25. Ghorbani, M., Lashkenari, M. S., and Eisazadeh, H. Application of polyaniline nanocomposite coated on rice husk ash for removal of Hg (II) from aqueous media. *Synthetic Metals*, **161**, 1430–1433 (2011).
26. Li, Z., Sun, X., Luo, J., and Hwang, J. Y. Unburned carbon from fly ash for mercury adsorption: II. adsorption isotherms and mechanisms. *Journal of Minerals and Materials Characterization and Engineering*, **1**(2), 79–96 (2002).
27. Rio, S. and Delebarre, A. Removal of mercury in aqueous solution by fluidized bed plant fly ash. *Fuel*, **82**, 153–159 (2003).
28. Somerset, V., Petrik, L., and Iwuoh, E. Alkaline hydrothermal conversion of fly ash precipitates into zeolites 3: The removal of mercury and lead ions from wastewater. *Journal of Environmental Management*, **87**, 125–131 (2008).
29. Danwanichakul, P. and Danwanichakul, D. Mass Transfer Analysis of Mercury (II) Removal from Contaminated Water by Non-Porous Waste Tire Granules. *European Journal of Scientific Research*, **36**(3), 363–375 (2009).
30. Kadirvelu, K., Kavipriya, M., Karthika, C., Vennilamani, N., and Pattabhi, S. Mercury (II) adsorption by activated carbon made from sago waste. *Carbon*, **42**, 745–752 (2004).
31. Wahi, R., Ngaini, Z., and Jok, V. U. Removal of Mercury, Lead and Copper from Aqueous Solution by Activated Carbon of Palm Oil Empty Fruit Bunch. *World Applied Sciences Journal*, **5** (Special Issue for Environment), 84–91 (2009).
32. Silva, H. S., Ruiz, S. V., Granados, D. L., and Santángelo, J. M. Adsorption of Mercury (II) from Liquid Solutions Using Modified Activated Carbons. *Materials Research*, **13**(2), 129–134 (2010).
33. Zabihi, M., Asl, A. H., and Ahmadpour, A. Studies on adsorption of mercury from aqueous solution on activated carbons prepared from walnut shell. *Journal of Hazardous Materials*, **174**, 251–256 (2010).
34. Tan, Z., Qiu, J., Zeng, H., Liu, H., and Xiang, J. Removal of elemental mercury by bamboo charcoal impregnated with H_2O_2. *Fuel*, **90**, 1471–1475 (2011).
35. Pulido, L. L., Hata, T., Ishihara, Y. I. S., and Kajimoto, T.. Removal of mercury and other metals by carbonized wood powder from aqueous solutions of their salts. *Journal of Wood Science*, **44**, 237–243 (1998).
36. Wahby, A., Abdelouahab-Reddam, Z., El Mail, R., Stitou, M., Silvestre-Albero, J., Sepúlveda-Escribano, A., and Rodríguez-Reinoso, F. Mercury removal from aqueous solution by adsorption on activated carbons prepared from olive stones. *Adsorption*, **17**, 603–609 (2011).
37. El-Shafey, E. I. and Al-Haddabi, K. N. Reductive sorption of heavy metals from aqueous solutions on a carbonaceous sorbent chemically prepared from date palm leaflets. *AIEEE*, 978-1-4244-5089-3/11/©2011 IEEE (2011).
38. Lohani, M. B., Singh, A., Rupainwar, D. C., and Dhar, D. N. Studies on efficiency of guava (*Psidium guajava*) bark as bioadsorbent for removal of Hg (II) from aqueous solutions. *Journal of Hazardous Materials*, **159**, 626–629 (2008).

12 Some Physicochemical Measurements of Chitosan/Starch Polymers in Acetic Acid-Water Mixtures

Virpal Singh and Kamlesh Kumari

CONTENTS

12.1 INTRODUCTION

Chitosan is a modified natural carbohydrate polymer prepared by the partial *N*-deacetylation of chitin. Chitin is an amino polysaccharides (combination of sugar and protein) an abundantly available natural biopolymer found in the exoskeletons of crustacean like shrimp, crabs, lobster, and other shellfish [1-3]. Chitosan is also found in some microorganisms like yeast and fungi. The primary unit in the chitin polymer is 2-deoxy-2-(acetylamino) glucose. These units are combined by ß-(1, 4) glucosidic linkages forming a long chain linear polymer. The degree of hydrolysis (deacetylation) has a significant effect on the solubility and rheological properties of the polymer. The structure of chitosan is shown in scheme 1.

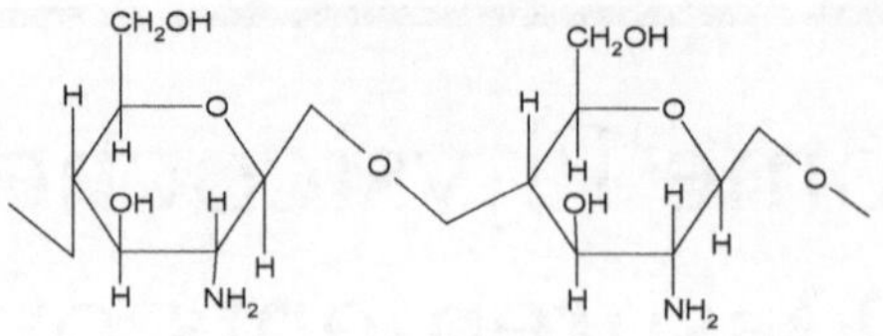

(1)

As a natural carbohydrate, starch is considered to be one of the major constituent of human diet. It is biodegradable and naturally metabolized by human body. Starch is mainly composed of two homopolymers of *D*-glucose, amylose is mostly linear α-D (1, 4)-glucan and branched amylopectin, having the backbone structure as amylose but with the many α-1, 6 linked branch points. There are lots of hydroxyl groups on starch chains, two secondary hydroxyl groups at C-2 and C-3 of each glucose residue, as it is not linked. Linear and cross-linked structures of starch are shown in scheme 2. Evidently starch is hydrophilic. The available hydroxyl group on the starch chains potentially exhibit reactivity specific for alcohols. In other words, they can be oxidized and reduced, and may participate in the formation of hydrogen bonds, ethers. Starch has different proportions of amylose and amylopectin ranging from 10–20% amylose and 80–90% amylopectin depending on source [4-6]. Starch occurs naturally as discrete granules since the short branched amylopectin chains are able to form helical structures which crystallize. Starch granules exhibit hydrophilic properties and strong inter–molecular association via hydrogen bonding formed by hydroxyl groups on the granule surface [7-10]. It is important to know the physicochemical properties of chitosan/starch solutions having different weight ratios.

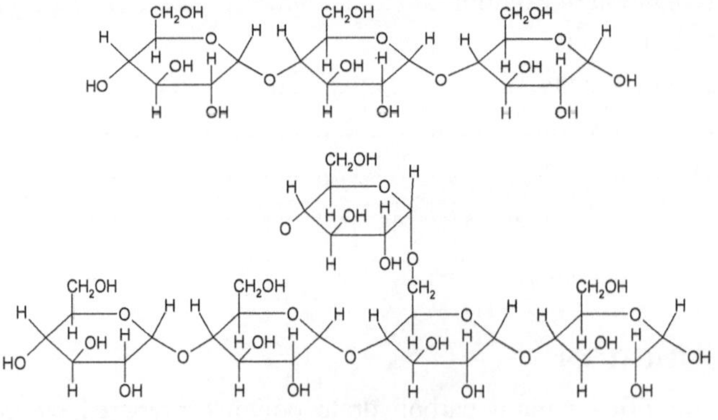

(2)

12.2 EXPERIMENTAL

12.2.1 Materials

Chitosan is supplied by Fluka Chemie (Germany) and starch is procured from Himedia (India). Acetic acid (99.5%) is purchased from Merck (Germany). For the preparation of solutions, double distilled water is used.

12.2.2 Preparation of Solution

The solutions of chitosan/starch (ch/st) are prepared in different weight ratios (90/10, 80/20, 70/30, 60/40, 50/50, 40/60) using a balance with a precision of 10^{-7} kg. To prepare the solution of chitosan it is dissolved in acetic acid (1, 2, 3, and 4%) at room temperature that is 25°C while stirring for three hours. Starch solution is prepared by dissolved it in water at 95°C while stirring for 20 min followed by cooling. Then both the solutions are mixed together and kept for 24 hr which resulted in bubble free clear solution.

12.2.3 Measurements

Viscosity of chitosan, starch, chitosan/starch blends in different weight ratios is measured with the help of Brookfield Digital viscometer, modal Dv-E version 1.00. Viscosity is a measure of fluid resistance to flow. The principle of operation of the DV-I is to drive a spindle (which is immersed in test fluid) through a calibrated spring. The viscous drag of the fluid against is measured by spring deflection. The viscometer is calibrated by using viscosity standard fluid and accuracy was ±1%. The speed of spindle is fixed at 30, 50, 60, 100 rpm and spindle number (1, 2, 3, 4, 5, 6, and 7) is also fixed according to nature of the solution. The spindle number 1 is for very low viscous solution and spindle number 7 is used for very high viscous solution.

Density is measured with the help of pychnometer having a bulb volume of 10 cm^3 and a capillary bore with an internal diameter of 1mm. The pycnometer was filled to specific volume with accuracy followed by the measurement of mass of solution.

The measurement of refractive index of 1 g of polymeric sample is done by Abbes refractometer. The scale is adjusted so that the boundary between light and dark coincides with the center of the cross hairs. The refractive index on the top scale in the lower part of the viewer is read and recorded.

12.3 DISCUSSION AND RESULTS

The results are plotted in terms of apparent viscosity versus concentration of ch (g)/st (g) solution in different compositions. Figure 1 represents the variation in viscosity of solution with concentration of chitosan. The concentration of acetic acid is varied from 1 to 4%. It is observed that the viscosity of blended solution increases with an increase in concentration of chitosan in solution and decreases with an increase the speed of the rotation. This is due to the fact that chitosan is a high molecular weight polymer as compared to starch and a small increase in the mass of this polymer results in more viscous polymer solution. The decrease in viscosity with increase in the rpm of spindle is due to the enhance rate of shear stress. As the speed of the spindle increases shear stress increases which results in an increase in shear rate. The density of different solution is also measured at room temperature (i.e. 25°C) and represented in Figure 2. It is noticed that there is an appreciable increase in density of polymer with increase in concentration of solvent as well as concentration of chitosan. The Figure 3 shows that the refractive index of solution increases with increasing % of acetic acid.

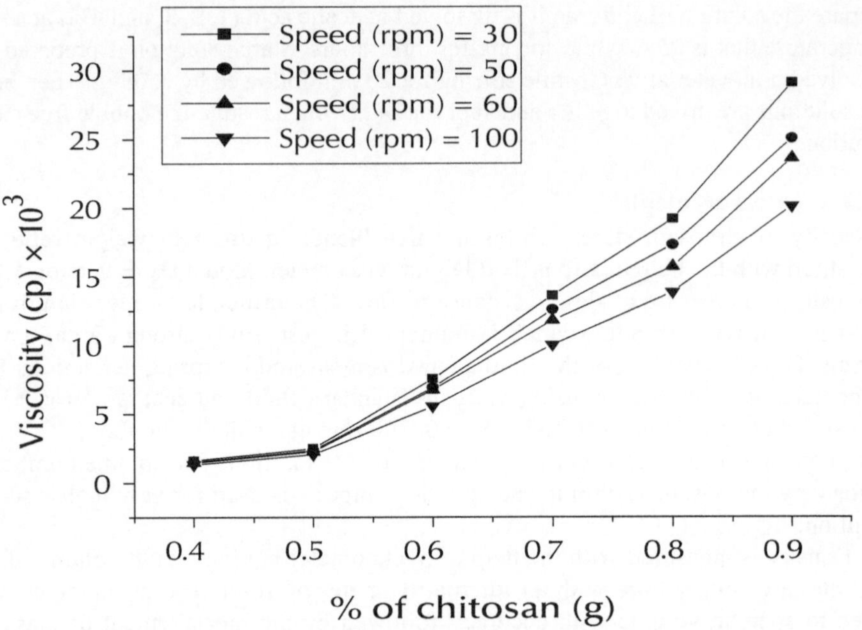

FIGURE 1 Variation in viscosity of polymer solution with polymer concentration in 2% of acetic acid solution and spindle speed.

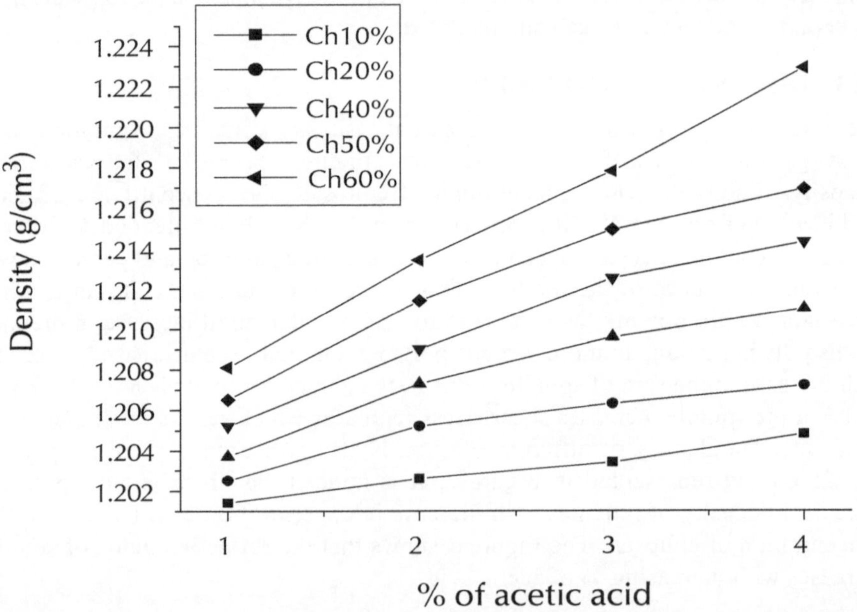

FIGURE 2 Dependence of density of solution on chitosan concentration and percentage of acetic acid solution.

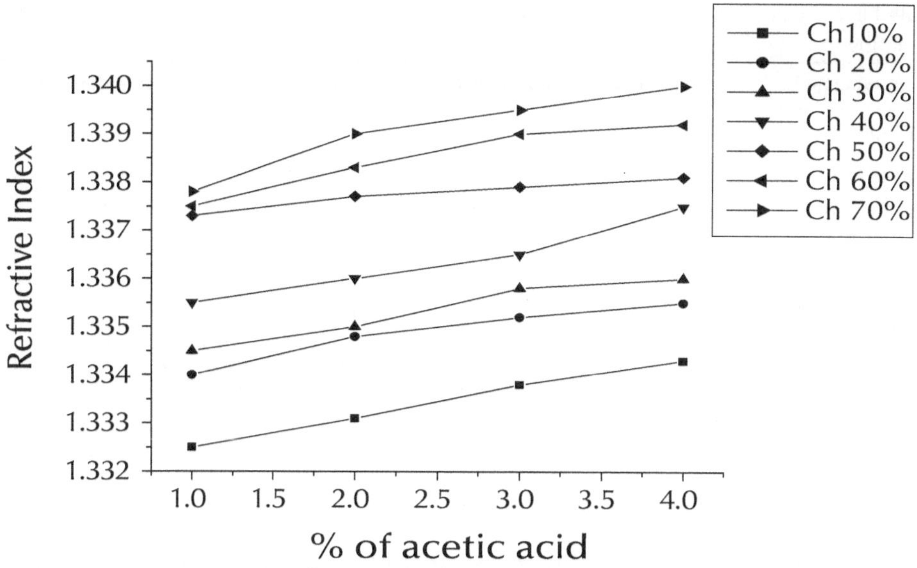

FIGURE 3 Dependence of refractive index of polymer solution on chitosan concentration and percentage of acetic acid solution.

Steady Shear Viscosity—In order to describe the steady-shear rheological properties of samples, the data was fitted to the well known power law model.

$\sigma = K \dot{y}^n$

where σ = shear stress; Pa; K = consistency coefficient; Pa sec; $\dot{y}$ = shear rate; sec^{-1}; n = flow behavior index, dimensionless.

Apparent viscosity can be expressed as

$\eta = \sigma/\dot{y}$

In applying the Mitschka method, the flow behavior index can be measured from the slope of the logarithm of shear stress verses logarithm of rotational speed plot.

$n = d(\log_{10} \sigma)/d(\log_{10} N)$

where σ = shear stress, Pa, N = rotational speed rpm.

The flow behavior index (n) and consistency coefficient are obtained by plotting graphs between log (shear stress) and log (shear rate) for different solutions prepared in the predefined compositions. The obtained values of consistency coefficient and flow behavior index are given in Table 1. The value of n for 2% acetic acid (chi/st as 0.4/0.6) and 4% acetic acid (ch/st as 0.4/0.6) are found to be 0.595 and 0.551. These values of n indicates that the solutions are both highly pseudo plastic in nature and this may be assigned to deviation from Newtonian behavior due the presence of entanglements. It is observed that all the blends compositions exhibit non Newtonian behavior with increasing chitosan concentration. Rheograms of all the blends are found to lay between the rheograms of pure chitosan and pure starch components over the entire range.

TABLE 1 Values of flow behavior index (n) and consistency coefficient in 2% acetic acid solution.

Ch (g)/st (g)	Flow behavior index (n)	Log (consistency coefficient) = log (K)
0.4/0.6	0.595	0.511
0.5/0.5	0.521	0.847
0.6/0.4	0.588	1.721
0.7/0.3	0.742	1.199
0.8/0.2	0.764	1.2180
0.9/0.1	0.701	1.5910

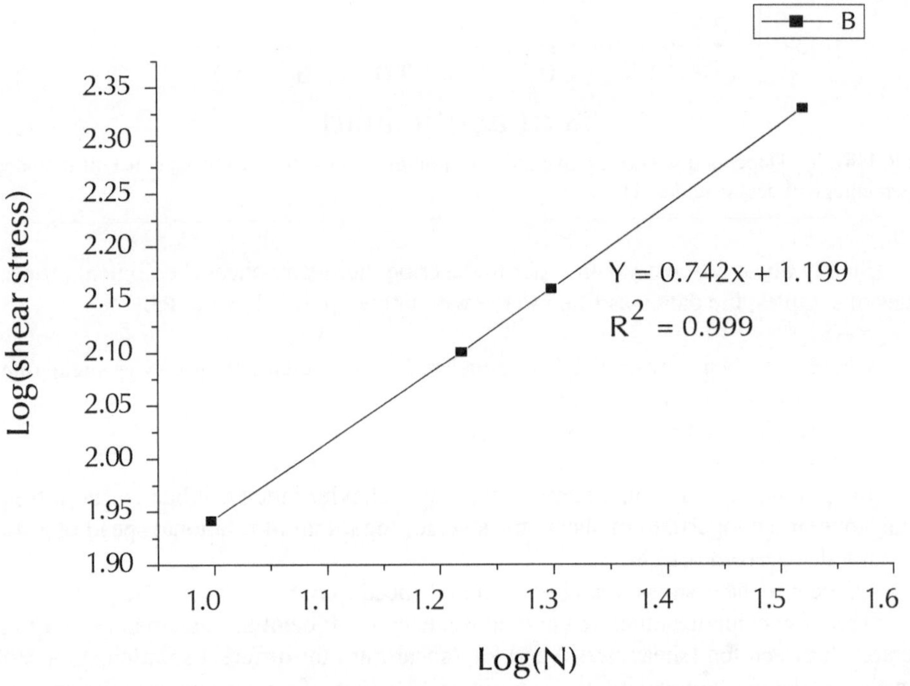

FIGURE 4 Influence of spindle speed on shear stress for 2% acetic acid solution of ch/st (0.7/0.3).

Figure 5 Influence of spindle speed on shear stress of solution having different concentrations at 25°C. One can see that, for 2% acetic acid solution, as the mass of chitosan increases in ch/st solution the shear stress increases. The speed of the spindle increases shear stress which further affects the shear rate.

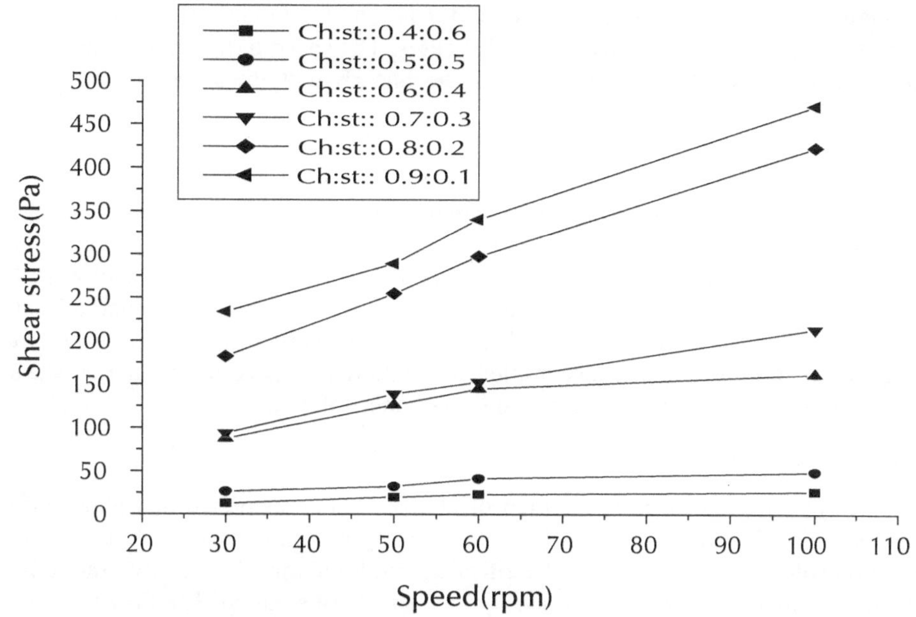

FIGURE 5 Influence of spindle speed on shear stress of solution having different concentrations at 25°C.

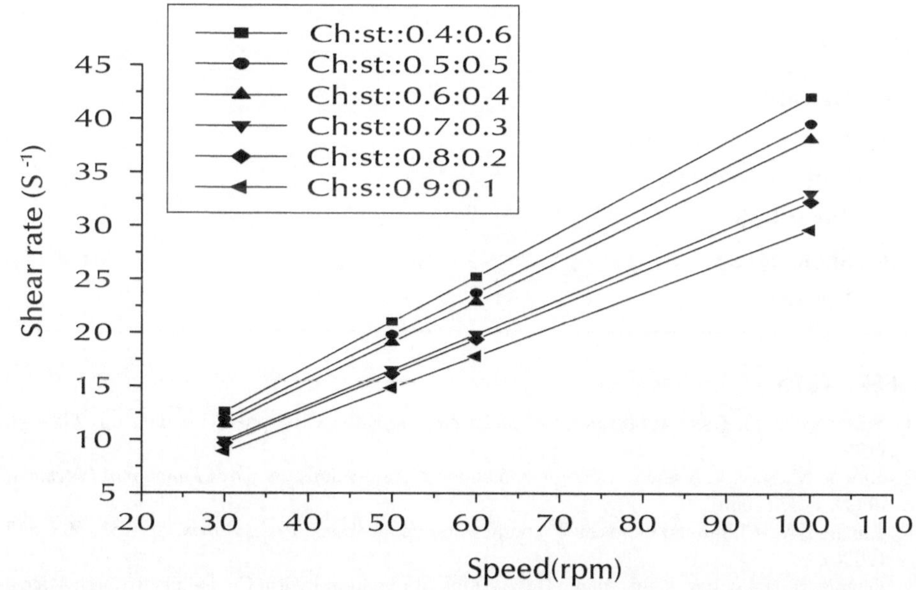

FIGURE 6 Influence of spindle speed and solution concentration on shear rate at 25°C.

Figure 6 shows that for 2% acetic acid solution, as the amount of chitosan increases in ch/st solution, the shear rate decreases. The speed of the spindle increases the shear rate on solution. Shear rate increases linearly with the speed of the spindle for the different ch/st solution.

12.4 SUMMARY

Physicochemical properties of chitosan/starch solution have been studied. Chitosan/starch solutions of different concentrations (90/10, 80/20, 70/30, 60/40, 50/50, 40/60, 30/70, 20/80, 10/90) are prepared in dilute acetic acid solution (1, 2, 3, and 4%). It is observed that the solution of chitosan/starch is miscible over entire range of concentration. The solution properties such as viscosity, density, and refractive index are measured. The influence of concentration of solution and speed of rotation on shear stress and shear rate are also determined for polymer solutions.

12.5 CONCLUSION

This study concluded that the solution of chitosan/starch exhibited non Newtonian flow behavior. It is observed that the small quantity of chitosan produced enormous effect on solution viscous. Thus, the effect caused by the polymer concentration resulted in an increase in apparent viscosity and density of solution. The flow behavior index, n increases when chitosan concentration is increased. It is also observed that the shear stress and shear rate increase linearly with speed of rotation and concentration of chitosan. The finding of this study may be useful for several industrial processes such as mixing and fluid transport.

KEYWORDS

- **Density**
- **Flow Behavior Index**
- **Refractive Index**
- **Shear Rate**
- **Shear Stress**
- **Viscosity**

REFERENCES

1. Salamone, J. C. *Concise polymeric material encyclopedia*. CRC press, London, pp. 238–246 (1999).
2. Kumar, M. N. V. R. A review of chitin and chitosan applications. *Reactive Functional Polymers*, **46**(1), 1–27 (2000).
3. Rinaudo, M. Chitin and chitosan: Properties and applications. *Prog. Polym. Sci.*, **31**, 603–632 (2006).
4. Ghania, H. and Gilles, P. Enzymatic Degradation of Epichlorohydrin Cross-linked Starch Microspheres by α-Amylase. *Pharmaceutical Research*, **16**(6), 867–875 (1999).
5. Baran, T. N., Mano, F. J., and Reis, L. R. Starch-chitosan hydrogels prepared by reductive alkylation cross-linking, *Journal of material Science: Material medicine*, **15**, 759–765 (2004).

6. Atyabi, F., Manoochehri, S., Moghadam, S. H., and Dinarvand, R. Cross-linked starch microspheres: Effect of cross-linking condition on the microsphere characteristics. *Arch Pharm Res*, **29**(12), 1179–1186 (2006).
7. Elvira, C., Mano, J. F., San Roman, J., and Reis, R. L. Starch Based Biodegradable Hydrogels with potential Biomedical Applications as Drug Delivery Systems. *Biomaterials*, **23**, 1955–1966 (2002).
8. Kurkuri, M. D., Kulkarni, A. R., and Aminabhavi, M. Some physicochemical measurements of chitosan polymer in acetic acid-water mixtures at different temperatures. *Journal of Applied Polymer Science*, **86**, 526–529 (2002).
9. Hülya, Y. and Ceyhun, B. Preparation and Biodegradation of Starch/Poly caprolactone Films. *Journal of Polymers and the Environment*, **11**(3), 107–113 (2003).
10. Maity, S.. "*Analysis and characterization of polymers*" Anusandhan Prakashan Midnapore, pp. 85–94 (2003).

13 Morphology-mechanical Properties Correlation in Polymer Composites with Bamboo Fibers

Rameshwar Adhikari

CONTENTS

13.1 INTRODUCTION

Bamboos, the flora belonging to perennial grass, are regarded as fastest growing plants on the earth. They grow almost everywhere without much human effort, protect environment, and produce one of the strongest natural fibers that have been accompanying the human lives from times immemorial. They not only provide foods, medicines, and habitats to several living beings but also are strong candidates for applications as fillers in plastics composites and for several high-tech applications. The purpose of this chapter is to review the relationship between structure and mechanical properties of the composites comprising bamboo fibers as fillers. The effect of different surface treatments and modification as well as compatibilization of the fibers on morphology and mechanical behavior of the composites will be addressed. In particular, the structure property correlation in completely biodegradable polymer composites will be highlighted. A short introduction to the morphology and the properties of the bamboo fibers will be followed by a review of morphology of the bamboo fiber filled polymer composites comprising thermo sets, thermoplastics, biodegradable polymers, and polymer blends as matrix. We will attempt to critically review the mechanical performance of the polymer/bamboo composites with special reference to the fracture surface characterization of the composites by electron microscopy.

During the last century, plastics materials have emerged as an essential element of human life. Almost every activity of our life is directly or indirectly influenced by their applications. These have made us so dependent that it appears impossible to think about sustaining our contemporary form of society without plastics. Indeed, these materials have already occupied central role in various areas of materials science and engineering, health sciences, nanotechnology, electronics, energy concerns, and even in tackling with several environmental issues. On the other hand, the plastics wastes have become today big environmental problems worldwide. The challenges are not limited to human health alone but are growing as potential risks to the lives of castles, microorganisms, sea animals threatening simultaneously the growth of several floras. The matter that then emerged as God is rapidly turning into an *Evil*. And, neglecting these severe global environmental issues and keeping our eyes closed towards the fact of heavily increasing oil prices from which the raw materials for most of the plastics are derived, we are producing more and more synthetic polymers and selling them to transform our efforts and time to money. Even in the domain of synthetic plastics, demands on light composite materials having high load bearing ability (strength), large deformability, and resistance to failure (toughness) coupled with different functional properties are increasing. For these purposes, the attraction towards the use of natural fibers as reinforcing fillers is in rise. It has been deeply realized today that we cannot avoid plastics and we, however, need to develop new plastics on the basis of naturally occurring biomass [1-6]. Several models of such plastics materials have been developed and introduced into the markets.

A great majority of natural materials are stronger and stiffer in fibrous form than in their bulk state. The natural fibers (such as jute, sisal, silk etc.) have gained importance as fillers in plastics because of their low cost, low density, rough surface, and reduced wear of processing machinery. However, full commercial potential of the natural fibers as reinforcing filler in plastics has not been achieved due to lack of research and development of high technology applications. One interesting area of research that is emerging as an important trend of contemporary materials science and technology comprises the development of completely biodegradable, environmentally benign composite materials. Many efforts have been made in the last decade to develop completely green composite materials whereby the reinforcing fillers comprise wastes and byproducts from carpentries, agriculture, and food industries [1, 7-12], [13] Among different kinds of fibers, those from bamboos are well known for their low density, excellent mechanical strength [14-16]. Besides their classical uses as construction materials and as a composite component to make houses and household articles, the bamboos have been utilized in several modern applications such as in concrete reinforcement [17], making sandwitch composites with metals [18]. It was shown recently that in the thermoplastic polyblend resins, the combination of bamboo structures into the aluminum laminate significantly increases the flexural and compressive properties of the laminates [17]. Efforts have been made even to mimic the structures of bamboo to design tough, strong, and still flexible nanoporous silicon using conducting polymers as a supporting material [19]. Bamboo/epoxy composites have been utilized to prepare special materials such as wood ceramics [20], footwear [21], and wind-mill turbines [22]. It was concluded that both sintering temperature and epoxy resin content have significant effects on the basic properties of the ceramics.

Keeping those new developments in mind, the aim of this chapter is to review the recent advances in polymer/bamboo fiber composites emphasizing the morphology formation and deformation behavior of the composites. In particular, the results from electron microscopy and tensile testing will be highlighted.

13.2 BAMBOO, FIBERS, AND NANOFIBERS: STRUCTURE AND PROPERTIES

From the view point of material science, bamboo can be regarded as a highly filled composite material comprising cellulose microfibrils as the filler embedded in the matrix comprising lignin, hemicelluloses and so on acting as matrix. It has an interesting macro and microstructures organized in hierarchical features which contribute to its structural integrity [14, 16, 23-26]. The mechanical properties of bamboos are highly anisotropic with strongly oriented molecules and fibers along the growth direction, the strength along longitudinal direction being, for instance, more than ten times, than in the transverse direction [14]. As a result, bamboo belongs to the class of viscoelastic materials like other polymers but cannot be sufficiently explained by the common cellular model of linear viscoelasticity. To study the morphology and different physical properties of the individual fibers, several chemical and mechanical methods have been developed [16, 27].

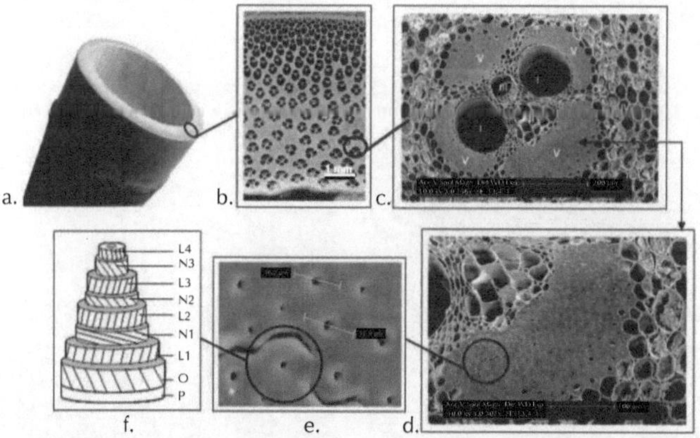

FIGURE 1 Bamboo microstructure: (a) bamboo culm, (b) cross-section of the culm showing the fiber distribution through the wall thickness, (c) vascular bundle, the main anatomical constituent of the plant, composed of vessels I, phloem II, protoxylem III, parenchyma tissue IV, and fiber bundles V, (d) bamboo fiber bundles composed of several elementary fibers, (e) elementary fibers, and (f) model of polylamellar structure of a thick-walled elementary fiber proposed by Liese [25] with thick lamellae (L1–L4), thin lamellae (N1–N3), primary wall (P) and outer sheet of secondary wall (O) [16]).

Several researchers have explored the hierarchical structure and morphology of various kinds of bamboos using different techniques ranging from optical microscopy, electron microscopy, and spectroscopic techniques. The details of the bamboo structure and properties may vary from species to species and depending upon the testing conditions, age of the culms as well as the way how the fibers are extracted. Nevertheless, the basic features are similar. Thus, in the following paragraphs, we will not differentiate between the bamboo types with respect to their origin, taxonomy, and preparation history. An overview of the bamboo structure is summarized in Figure 1 [16].

Bamboo fibers have the interesting combination of mechanical properties low density and high mechanical strength, which are most favorable among the well known fibers found in nature. This is attributable to the special hierarchical morphology of the bamboo shoot. The hollow bamboo culm (Figure 1(a)) comprises the highly oriented fiber bundles (the microfibrils) embedded in the matrix of lignin which are distributed more densely towards the outer surface (Figure 1(b)), the concentration of the fibers also increases as one moves away from the base of the plant. The elementary fibers in such a bundle consist of alternating thick and thin layers with different fibril orientation (Figure 1(f);[25]). According to Liese model [25], the fibers and parenchyma cell walls of bamboo have a multilayered structure arising from the alternation in the orientation of cellulose microfibrils which results in a structure with high tensile strength. In the thick lamellae, the fibrils are oriented at a small angle to the fiber axis, whereas the thin ones, containing higher amount of lignin than the thicker ones, show predominantly transverse orientation of the fibrils. This hierarchical polylaminate wall structure of the fibrils leads to an extremely high tensile strength of the culm [28].

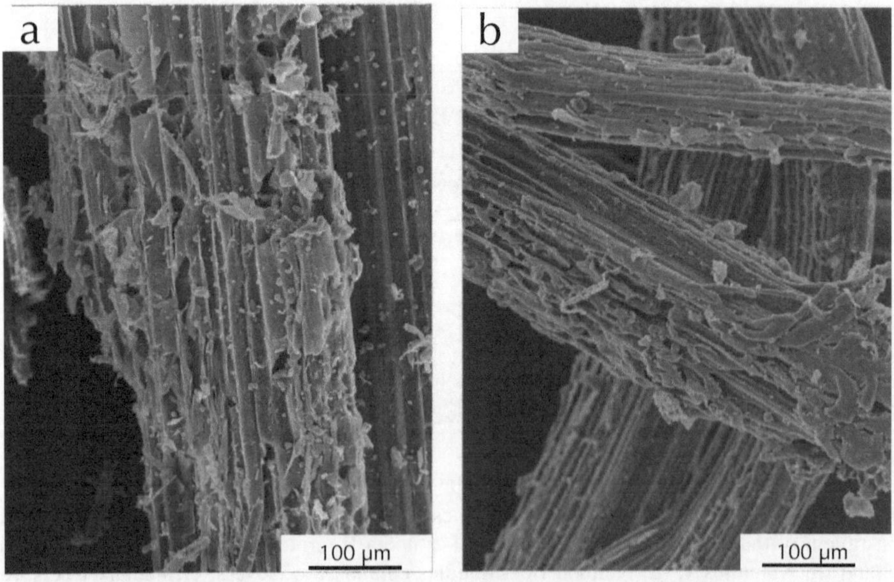

FIGURE 2 SEM micrographs of bamboo flour comprising bamboo fibers formed by mechanical crushing; (a) neat fiber from dried flour and (b) fibers after caustic soda treatment.

The detailed structure of the bamboo micro and nanofibers has been investigated by electron microscopy. In particular, scanning electron microscopy (SEM) using secondary electron imaging has been employed [12, 14, 16, 28-30]. Occasionally, transmission electron microcopy (TEM) and atomic force microscopy (AFM) are used to study the detailed nanostructures of the fibrils and fibers [30-34]. Figure 2 presents SEM micrographs of technical bamboo the flour prepared by mechanical crushing [12]. In the neat flour (Figure 2(a)), the bundles of long microfibrils, several μm thick and glued together in a bunch can be seen while the fibrils are separated and exposed out due to the dissolution of alkali soluble parts such as hemicelluloses in the treated flour (Figure 2(b)). Each fibril is in fact sclerenchymal cell 10-20 μm thick separated by a wall 1-5 μm in thickness [35], which itself comprises many continuous and twisted cellulose nanofibers (CNF) elongated along the culm axis. The fibers are staggered in such a fashion that they assume the shape of metal cables formed by several twisted wires. This unique architecture of the bamboo fibrils endows the bamboos with special mechanical properties.

FIGURE 3 TEM images of different magnifications showing the morphology of bamboo CNFs; the specimens were stained with phosphotungstic acid to enhance the contrast [30].

In order to fully utilize the full potential of special organization of the cellulose fibers in bamboo, it would be essential to exfoliate them from the fibrils into individual nanofibers. The CNF are long thread-like bundles of cellulose molecules stabilized laterally by hydrogen bonds. Depending on their origin and processing conditions, CNFs have diameters in the range of a few nanometers to several tens of nanometers [30, 36], and their length can reach up to a few millimeters [30]. The researchers have used various methods to extract the CNFs from bamboo culms [16, 37]. One of the popular ways of fiber extraction is offered by steam explosion [30, 37-41] in which often the chemical and mechanical processes are combined to obtain high quality products with reasonable yield. The typical procedure to extract the fibers comprises the steam explosion process followed by action with acid, microorganisms, and high shear mechanical agitation and so on. [30]. The structure and the properties of the fibers obtained depend, obviously, on the type of the bamboo chosen and the processing routes selected for their extraction. An example of highly oriented fibers as imaged by TEM is presented in Figure 3.

In a recent study, ultra long CNF (more than a millimeter long and 30-80 nm thick) were extracted from bamboo shoots by using conventional chemical treatment combined with intense mechanical agitation and by freeze drying [30]. Each nanofiber was shown to consist of bundles of cellulose nanofibrils 1-5 nm thick and several μm long, the pure cellulose possessing crystallinity of about 60% and thermal stability until a temperature of 300°C.

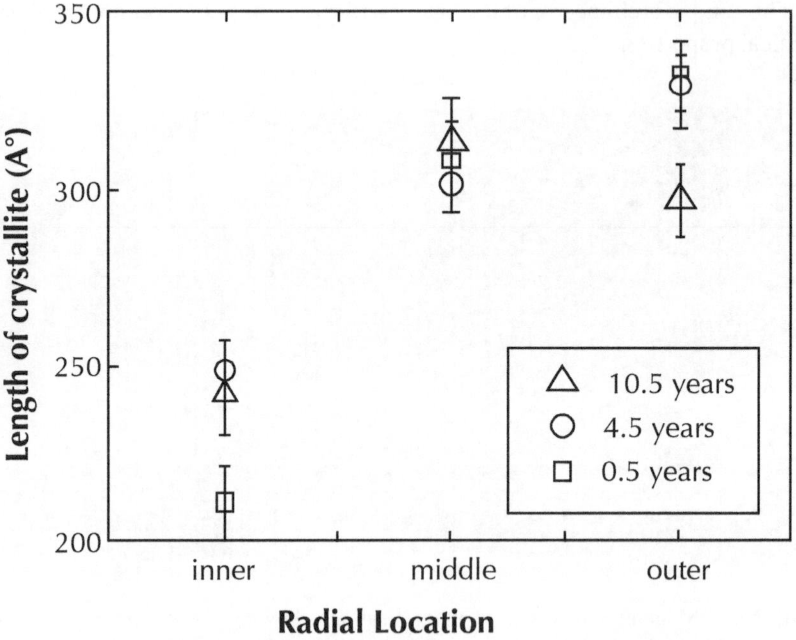

FIGURE 4 Length of the cellulose crystallites of bamboo at the ages of 0.5, 4.5, and 10.5 years. Error bars are based on the accuracy of the experiment and data analysis [42].

The bamboo fibers (technical fibers, microfibrils, and nanofibers) have been characterized by different spectroscopic methods such as electron spectroscopy for chemical analysis (ESCA), X-ray diffraction, nuclear magnetic resonance and infrared spectroscopy (such as [30, 42-46] to determine the chemical composition, crystalline texture, and crystallinity of the fibrous composite.

X-ray diffraction studies revealed that the crystal structure of the bamboo fibers is dependent on the nature of the bamboo type, age of the culm as well as the distance of the location from the center of the hollow culm. The length of the crystals, for instance was found to increase from ca. 20 nm to over 30 nm when moving from the inner wall of the bamboo to the outer wall (see Figure 4), similar trend being shown by the thickness of the crystals [42].

13.3 STRUCTURE PROPERTY CORRELATION IN BAMBOO/POLYMER COMPOSITES

The bamboo/polymer composites are reported with all classes of polymers-thermosets, thermoplastics, blends, and in combination with inorganic fillers, and biodegradable polymers. The goal has been to improve the polymer filler adhesion, design of specific properties, and study of mechanical properties in such studies. The fibers in various forms (variation in thickness and length of fibers) and under processing conditions have been used with and without compatibilizers of various origins-both synthetic and biobased [47-54]. The products thus fabricated have been targeted for short to long time applications.

13.3.1 Thermoplastics/Bamboo Composites

Thermoplastics such as polyethylene (PE), polypropylene (PP), different polyesters, and other semicrystalline polymers [53, 55-59] have been compounded with technical bamboo fibers and flour as well as microfibrillated cellulose (MFC) fibers from bamboo with the aim of reinforcing the matrix and also controlling crystallization behavior of the semicrystalline polymers, which are generally unable to undergo environmental degradation. Recently, many researchers are motivated to use also the CNF as reinforcing agents in different polymer matrices due to their high potential to enhance mechanical and several functional properties [3, 4, 8, 30, 32, 37, 60]).

Most of the studies have mentioned the alkali treatment of the natural fibers to expose cellulose fibers as pure components. It is generally claimed to have obtained better mechanical properties of the composites achieved by alkali treatment of the fibers. The researchers have focused particularly on the mechanical performance enhancement, control of crystalline morphology of the polymer matrix.

Several studies have demonstrated that the technical bamboo fibers can act in many cases as the nucleating sites for the growth of the spherulites [55, 56, 61]). For instance, in case of bamboo fiber (BF) filled isotactic polypropylene (iPP), the growth of individual spherulites was facilitated on the surface of each fiber also hindering to some extent the growth of three dimensional crystals (see Figure 5(a)). However, the story changes when the composite is prepared using compatibilizer enabling a strong bonding between hydrophobic polymer matrix and relatively hydrophilic cellulosic fibers (Figure 5(b)). The transcrystalline growth of spherulites on the surface of the bamboo fiber is sufficiently high when a small amount of maleic anhydride grafted polypropylene (MAPP) is used in the iPP/BF composite. It is interesting to note that the bamboo fiber nucleates more and more β phase in propylene endowing the polymer composite an additional benefit of increased toughness as β form of iPP comprising hexagonal β crystalline phase are known to be tougher than the monoclinic α phase [62].

FIGURE 5 Optical micrographs of iPP/bamboo composites: (a) with neat bamboo fibers and (b) composites containing MAPP; the magnifications in both the micrographs is 100 x [55].

The nucleation effect resulting in transcrystalline morphology of the composites is reflected also in melting behavior of the polymers. The effect of the bamboo fiber acting as β phase nucleating agent was also confirmed by differential scanning calorimetric analysis. Later Kori et al. [61] studied bamboo fiber reinforced composites of polybutylene succinate (PBS) with respect to the evolution of crystallization and viscoelastic behavior as a function of processing temperature [61]. In this case, however, no nucleating effect of the bamboo fibers was observed. The spherulites began to grow from the bulk polymer matrix and not from the filler surface.

Regarding the performance of the bamboo fibers reinforced composite materials; they are similar to that of classical wood flour reinforced plastics [56]: there is an increase in the elasticity modulus with occasional improvement in the strength of the final products. As in wood flour reinforced plastics, the composites show brittle behavior as these materials are strongly crack sensitive.

The studies related to bamboo flour reinforcement of polymers have been extended to polymer blends in order to prepare hybrid composites such as that of PP and polylactic acid (PLA), which represents more environmentally friendly composites materials as the PLA part is biodegradable. Both the dispersion and resulting mechanical properties of PP/PLA/BF hybrid composites are found to increase with the fiber matrix adhesion offered by the addition of MAPP [58].

13.3.2 Bamboo Fiber Composites with Thermosetting Resin and Rubber

Bamboo fibers as filler have been appreciated greatly for the reinforcement of thermo sets, in particular in epoxy and polyester based resins due to the technical importance of the letters. Bamboo fibers and mats reinforced epoxy resins comprising up to 65% weight fraction of filler can be conveniently prepared which exhibit superior properties than their other natural fiber reinforced counterparts [28]. Also, the treatment of the fibers by various agents influences the matrix/filler adhesion and mechanical properties of the composites. Comparative study of the epoxy resins with silane and alkali treated bamboo fibers show that the silane treatment does not bring any noticeable improvement in the tensile mechanical properties. The alkali treatment leads to the removal of waxy materials and amorphous hemicelluloses and significant improvement in mechanical properties [63-65]). The fibers cannot be however mixed to the epoxy resin to all compositions; there is an optimum fiber concentration as the fiber content also increases the void formation tendency of the composite materials [63]. Interestingly, the flexural and compressive mechanical properties are found to improve considerably with increasing amount of alkali treated bamboo fibers in epoxy/glass fiber/bamboo fiber hybrid composites [65].

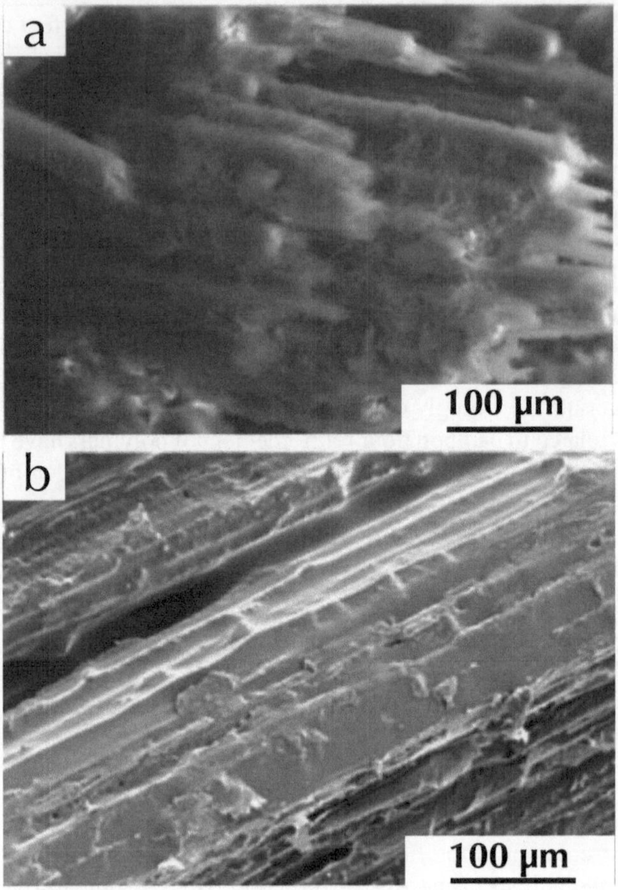

FIGURE 6 Scanning electron micrographs of the fracture surfaces of the bamboo fiber reinforced epoxy composites containing 48 volume % of the fiber: (a) composite with neat technical fiber and (b) composite with alkali treated technical fiber [16].

In a recent study, based on observation of various polymer/bamboo composites and comparing the results with the composites comprising other kinds of natural fibers, it has been further concluded that the reinforcing efficiency of the technical bamboo fibers can reach to that of glass fibers [16, 57]. Novel mechanical extraction process producing long bamboo fibers with excellent intrinsic mechanical properties and surface characteristics was developed [16]. In particular, the roughness of the fiber surface thus obtained has been attributed to the observed excellent wettability to the polymer matrix. Then, unidirectional long bamboo fiber/epoxy composites were successfully produced. The new studies revealed that alkali treatment is not indispensable to ensure the enhanced strength of the composite, as the untreated fibers already show reasonable adhesion with the epoxy matrix [16]. However, in most of the available literature works using the bamboo fibers obtained by other different methods claim that the mechanical properties of the composites (such as tensile strength, flexural strength, flexural modulus, toughness etc.) were optimized only on the treatment of the fibers with 5% caustic soda solution arising due to better compatibility of the fibers with the matrix resin [64, 66].

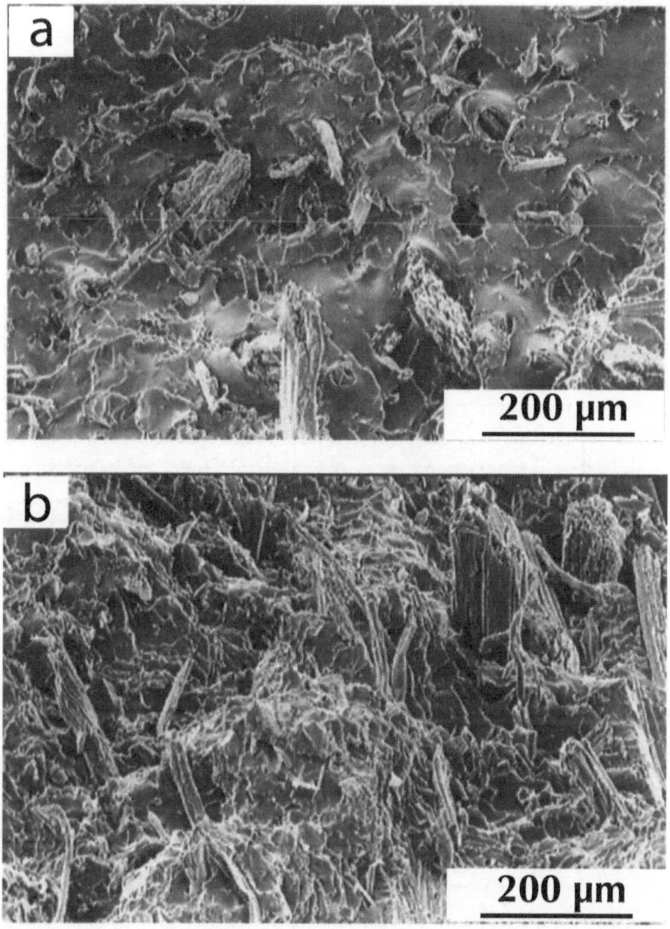

FIGURE 7 SEM images of tensile fracture surfaces of the natural rubber/bamboo fiber composites containing 10 phr of filler; (a) without coupling agent and (b) with coupling agent [47].

A typical morphology of the bamboo fiber reinforced epoxy resin has been exemplified in Figure 6. Even in the composite with neat fibers show a good adhesion with the matrix as the fibers remain intact and the matrix. Moreover, the matrix resin appears to fibrillate implying the plastic deformation of the matrix (Figure 6(a)). The surface of the fiber surfaces appear to be covered by the resin, implying that the failure occurs throughout the matrix. The SEM micrograph presented in Figure 6(b) shows the fracture surface of alkali treated bamboo fiber composite where the fibers are still covered by the polymer matrix. This observation suggests that matrix cohesive failure acts as the dominant failure mode in this composite. The properties of the bamboo fiber reinforced epoxy resins depend not only on the fiber treatment, thickness, and loading content but also on the fiber length. It has been found that the optimum level of mechanical strength of the composites can be obtained for fiber length of 10 mm [16].

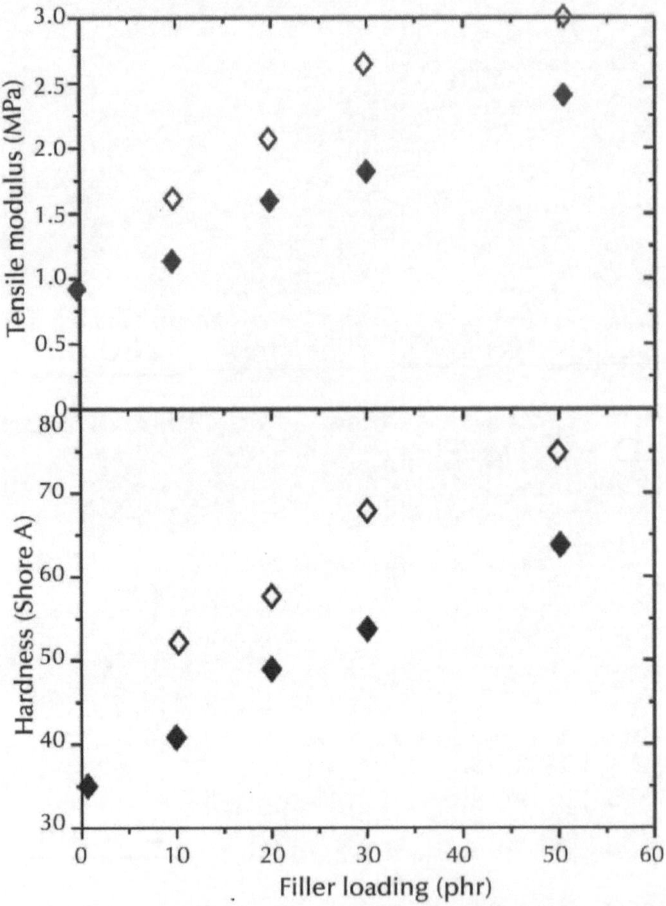

FIGURE 8 Variation of tensile modulus (a) and Shore A hardness (b) of natural rubber/BF composites; open symbol: with coupling agent and filled symbols without coupling agent [47].

Parallel to the reinforcement of the epoxy resins by bamboo fibers, several studies are devoted to the structure properties correlations in the polyester resin/bamboo composites [37, 67-69]). Bamboo fibers obtained *via* combined chemical and mechanical treatment were used to make unidirectional composites of polyester, in which the reasonable improvement in tensile properties was observed [37]. Also mercerization of the fibers with optimum alkali concentration was suggested as a means to attain the reinforcement of the highly filled composites [67].

Bamboo fibers have been used in different kinds of rubbers as reinforcing filler [13, 40, 47, 48]. Also in these cases, the effect of fiber loading, fiber treatment on the performance of the rubbers has been in the focus of the studies. The addition of bonding agents such as phenol formaldehyde and hexamethylene tetramine into the reaction mixture has been shown to dramatically enhance the mechanical performance of the rubber [47]. The use of silane coupling agent compensates the deterioration in dynamic properties by reducing at the same time the water absorption capacity of the composites.

Fracture surface morphology of alkali treated bamboo fibers reinforced natural rubber composites is presented in Figure 7 [47]. The fracture surface of the composite with untreated fibers exhibits the weak fiber/matrix interfacial adhesion (Figure 7(a)) as suggested by several holes and pulled out fibers from the fracture surface. The weak fiber/matrix compatibility of the composites is also demonstrated by large bunch (aggregates) of the fibers. Obviously, the fracture occurred at the matrix/fiber interface. In contrast, the fracture surface morphology of the rubber composite with coupling agent treated fibers shows feature of well developed interfacial interaction (Figure 7(b)). There is low pull out of fibers on the fracture surface. The surfaces of composites show that failure occurs at the fibers due to strong adhesion between bamboo fiber and rubber matrix (Figure 7(b)).

The effect of bonding agent is manifested in the mechanical properties of the composites as well (see Figure 8). With filler loading, as expected the mechanical properties such as tensile modulus (Figure 8(a)) and shore hardness (Figure 8(b)) both increase in the composites. The properties enhancements are significantly higher for the composites comprising coupling agent.

13.3.3 Green Polymer Composites

The progress in the use of natural fibers in the fabrication of composite materials has propelled the interest on the development of completely biodegradable or "green" composites. The last decade has seen enormous development in the scientific foundation as well as technical feasibility of the biocomposites which generally stand for the materials comprising completely biodegradable matrix and the fibers obtained from the renewable natural resources [2, 6-11, 60, 70]). Such composite have been often regarded as environmentally benign variable of the composites as these are believed to pose no or the least threats to the natural environment. The investigations on the green biocomposites using bamboo fibers as filler include polyhydroxybutyrate and copolymers [71], PLA [29, 52, 72], biodegradable polyesters and copolyesters [12, 72, 73] as matrix among others. All those studies have attempted to optimize the morphology properties correlations of the composites by designing filler parameters, coupling agents, blends formulations, and processing conditions of the biocomposites. The duration of uses of the biocomposites may range from a couple of days to couple of years depending upon the area in which these are employed.

In contrast to the properties frequently observed in the short fiber reinforced composite materials (i.e., a decrease in elongation at break, tensile strength, and impact strength), the compatibilized bamboo fiber composites of polyhydroxybutyrate-*co*-polyhdroxyvalerate (PHBV) showed a distinct improvement in tensile strength and impact toughness with unaltered elongation at break [71]. The SEM micrographs of fracture surfaces of the composites produced after impact testing are presented in Figure 9. The fracture surface morphology of the PHBV/bamboo pulp fiber composites without coupling agent (i.e., polymeric diphenylmethane diisocyanate, pMDI), significant fiber pull out, a process detrimental to the strength and modulus of the composites can be noticed (Figure 9(a)) indicating insufficient fiber bonding with the matrix polymer. With pMDI, interfacial bonding is strong, and hence, almost all the fibers are broken on the fracture surfaces (Figure 9(b)). The interfacial bonding of the components in the compatibilized composites and the resulting crazing mechanisms has been attributed for the observed simultaneous reinforcement and toughening effects [71].

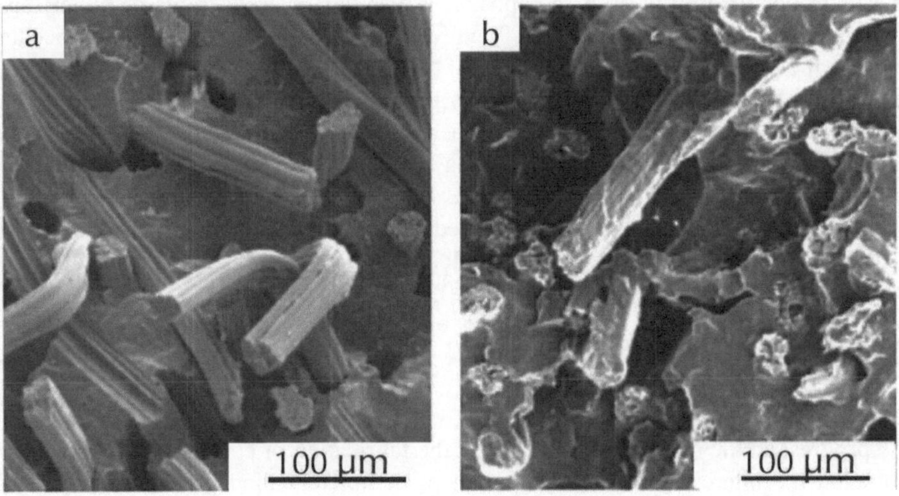

FIGURE 9 SEM micrographs of impact fracture surfaces of the PHBV/bamboo pulp fiber (80/20 w/w) composites: (a) without pMDI and (b) with pMDI [71].

Recently, similar studies were performed on polyhydroxybutyrate (PHB) filled with microfibrillated cellulose (MFC) from bamboo. It has been concluded that the MFC can effectively enhance the thermal stability and mechanical performance of the matrix polymer. It has been further pointed out that there is an optimum loading (10 wt. %) of the MFC beyond which the reinforcement does not exist.

In many studies, PLA was preferred as ecofriendly matrix for the preparation of bamboo fiber biocomposites [72, 29]. Likewise biodegradable polyesters and copolyesters [12, 50, 73] have been used as matrix to prepare new materials. Among the biodegradable polymers, PLA and PBS, aliphatic-aromatic copolyesters and so on have got particular commercial interest. Also with the advent of new biocomposite materials, there is an increasing trend of using biobased coupling and cross linking agents in their fabrication procedures.

The new kind of biobased coupling agent, lysince-diiocyanate (LDI), was recently utilized to prepare bamboo flour based composite materials with PLA and PBS, the hydrophobic polymers, as matrix [52]. It was observed that the isocyanate group (NCO) presents in the LDI plays an important role in the coupling efficiency of the agent in the bamboo fiber/polyester composite. There is an optimum concentration of NCO for the favorable activity of the coupling agent. Figure 10 shows an example of improvement of mechanical properties of the PBS/bamboo fiber composites. In absence of coupling agent, the tensile strength and elongation at break of the composite degrade with increase in fiber content. However, on processing the composites together with LDI, the tensile strength increases rapidly with filler loading; and the value of strain at break reaches a value much higher than that of uncompatibilized composites for all filler concentrations [52].

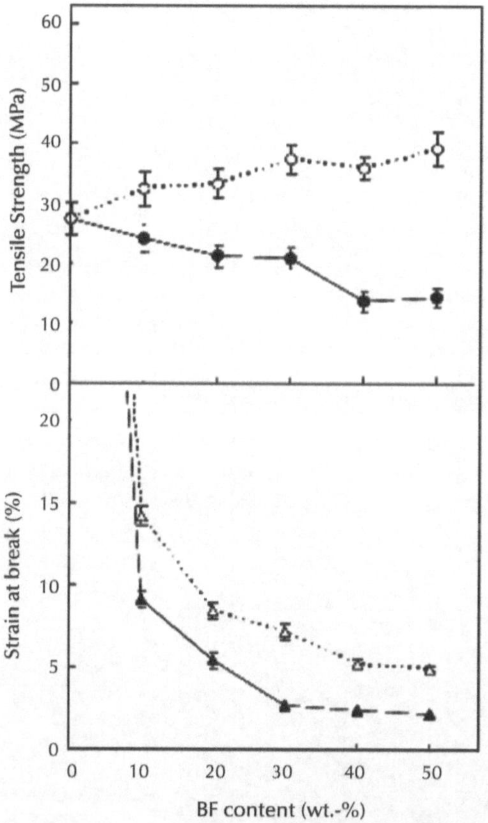

FIGURE 10 Effect of LDI coupling agent on the mechanical properties of PBS/BF composites at different filler concentrations: open symbols for compatibilized composites and filled ones for neat composites [52].

Adipic and terephthleic acids based aliphatic-aromatic copolyster introduced by the BASF SE is a commercially interesting biodegradable polymer which has been used in several studies to prepare composites and blends with other green components [7, 74-76] These studies further demonstrate the potential of completely green bio-compostes based on bio-erodible plastics and natural fibers.

The investigations have been extended to green composites using bamboo flour that may directly come from the carpentries without making (or with only little) modifications any so that the production cost is not high ([12]). The objective of the experiments was to develop highly filled compostable composite materials dispersing bamboo flour (not the fibers as discussed so far) in the matrix of biodegradable copolyester (Ecoflex, a commercial product of BASF, SE). The SEM micrographs of the compression molded composites with two different compositions are presented in Figure 11. The bamboo four prepared by mechanical crushing was mixed with polymer in an internal mixer prior to the molding.

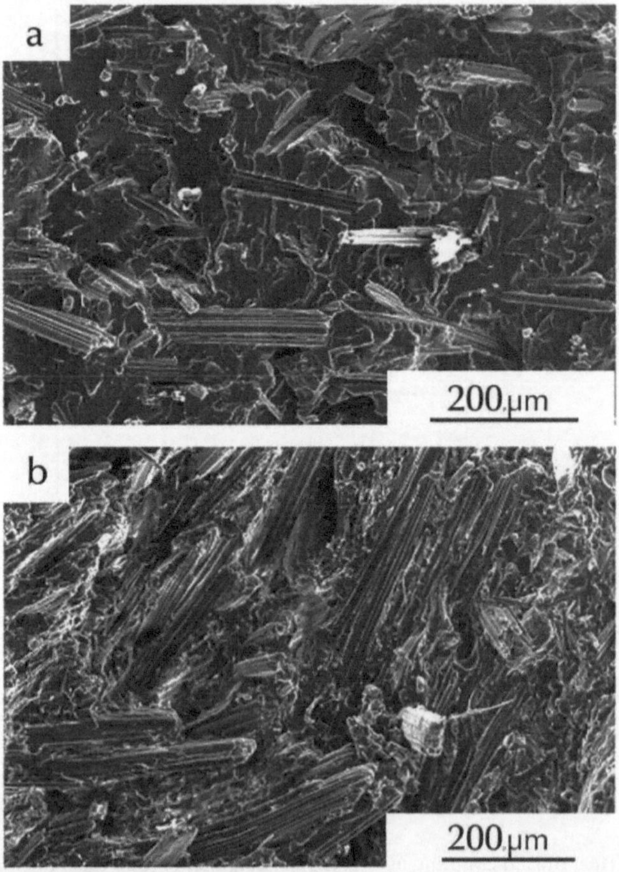

FIGURE 11 SEM images of Ecoflex/bamboo flour composites: (a) Ecoflex/20 wt. % bamboo flour and (b) Ecoflex/60 wt. % bamboo flour.

The micrographs in Figure 11 show that the bamboo flour can be successfully incorporated into the biodegradable polymeric matrix up to high content. At high filler concentration, the polymer acts merely as binding material. The filler is weekly adhered to the matrix as demonstrated by the pulling up of bamboo fibers. The filler dispersion is quite homogeneous (see Figure 11). Nevertheless, large amount of filler could be easily incorporated into the polymer matrix without any treatment of the fiber sand use of any interfacial agent. Tensile properties of the composite samples, however, deteriorated strongly beyond 40 wt. % of filler in spite of the alkali treatment of the fibers. The incompatibility of the bamboo flour with the Ecoflex matrix was also attested by X-ray diffraction and thermal methods ([12]).

Recently, polymer composites with cellulose nanofibers (CNF) as fillers have been shown to possess more interesting barrier properties coupled with excellent mechanical performance with very high bending strength, and stiffness (e.g., Young's modulus of about 150 GPa). The CNFs are available from various sources, such as cotton, flax, bamboo, jute, ramie, sisal, and wood. Their high aspect ratio and good dispersibility in the polymer matrix contribute to enhance the gas barrier properties of the composites, because the presence of the impermeable crystalline fibers can increase the travel path for gas leading to the slower diffusion processes. Thus, such composites are considered as promising candidates for preparing transparent packaging films with tailored barrier properties [3, 36, 40, 72].

There are a number of areas in which the cellulose nano and microfibers are used in the preparation of composite materials. The use of bamboo based fibrils as reinforcing agent in polymers is still growing [36, 40, 72]). In a recent study, bamboo nanofibers were extracted from bamboo pulp using novel hot compressed water (HCW) treatment which was then used as reinforcing agent in the polyurethane (PU) matrix [36].

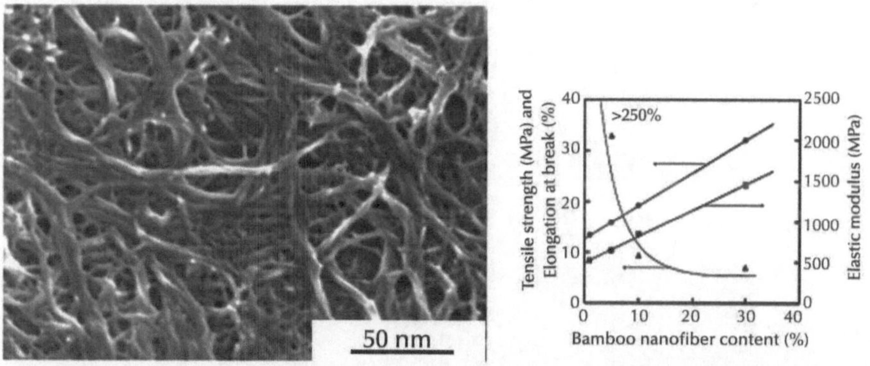

FIGURE 12 SEM images of CNFs extracted from bamboo culm through new HCW treatment (left) tensile properties of PU/bamboo CNF reinforced nanocomposites (right) [36].

The HCW treatment of the cellulose fibers followed by delignification procedure yields ultrafine nanocellulose fibers (see Figure 12, left) which upon incorporation into polymer matrix such as PU increases pronouncedly the strength and elasticity modulus of the composites (see Figure 12, right). The CNFs are found to act more effective nucleating agent in semicrystalline polymers. Moreover, microfibrillated cellulose (MFC) obtained from bamboo was recently used in combination with a biodegradable composites comprising PLA and technical bamboo fibers. It was found that the MFC plays a role of compatibilizer in these composites increasing both bending strength as well as fracture toughness via crack arresting [72].

13.3.4 Degradation of Polymer/Bamboo Fiber Composites

As biodegradability and sustainability have been the prime concerns in the developments of new environmentally benign polymer composites, several studies have been performed to examine the ease of biodegradation of the composites [52, 73-76]).

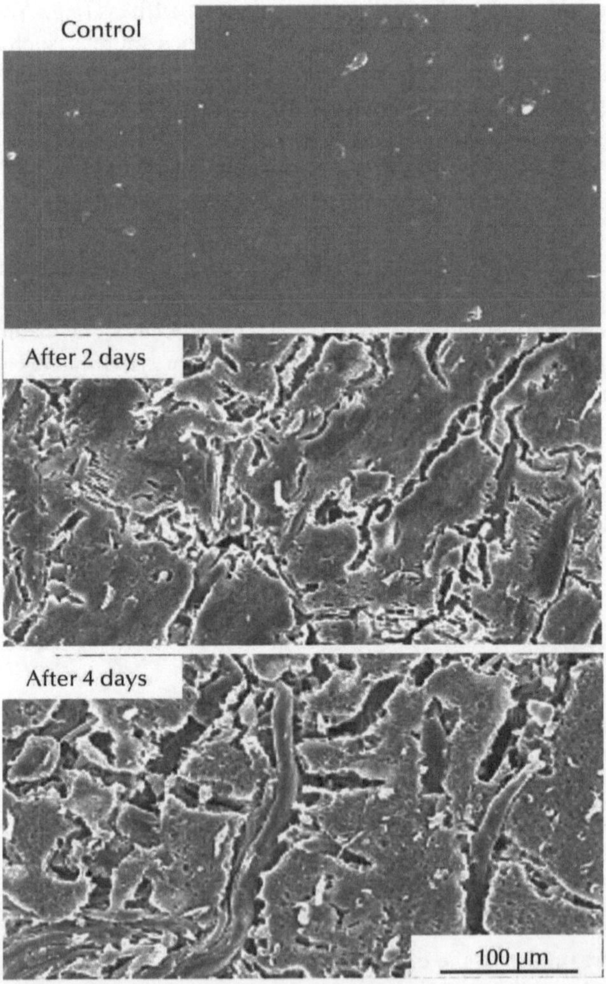

FIGURE 13 SEM micrograph of surfaces of compatibilized PBS/BF (70/30) composites after enzymatic degradation for different periods as indicated [52].

In polyesters based biocomposites, the degradation upon soil burial was observed after several weeks as evidenced by the weight loss of the samples and scanning electron micrographs of the fracture surfaces [52, 73]. As expected, the presence of the coupling agent slows down the biodegradation process of the composites as the filler/matrix integrity increases with the coupling of the composites components [73].

The accelerated biodegradation experiments can be simulated in the laboratory by treating the composite materials with enzyme solutions [52]. The enzymatic degradation of PBS/BF composites containing coupling agent LDI after different time intervals is depicted in Figure 13. The surface features of the composites changes dramatically even after a couple of days. Thus the composites of the kinds discussed in this section can be used for short time applications for example for making trays fast foods, and containers for fruits and vegetables, and so on. After the end-use, the chapter can be composted together with other degradable wastes that can be re-circulated into the natural environment.

13.4 CONCLUSION

The bamboo culm has been used in diverse areas ranging from making household articles to industrial applications. In addition to the classical uses of bamboos such as making toys, mats, chopsticks, furniture, flooring, music instruments, and so on. Bamboos, as one of the strongest nature made materials, have attracted considerable research interest also in the preparation of composites together with other materials. All sorts of plastics have been tested so far to fabricate composite materials with bamboo flours, fibers as well as micro and nanofibrils. A thorough survey of literature reveals that bamboo fibers, irrespective of their form and dimension, can offer significant reinforcement to the plastics as in the case of other natural fibers. The use of purified cellulosic fibers with proper compatibilization is a crucial aspect in the structure and properties of the composites. The intrinsic potential of the bamboo fibers as effective filler should be, however, still exploited. The deformation mechanisms in bamboo flour reinforcement of plastics is the least understood chapter which is perhaps acting as an obstacle in comprehensive understanding of complex structure properties correlation of bamboo fibers filled composites.

The advent of polymer/bamboo fiber composites has offered the application of these composites in various applications. Provided the issues related with chemical treatment of the fibers, economical ways of compatibilization of the compounds and optimization of process parameters well addressed, the bamboo fibers reinforced polymer composites can offer still more reliable service to the mankind in various ways. Those application areas may include sound proof devices, insulations, making toys, low weight bearing show cases and furniture, disposable trays and containers for food items, reinforcements for rubber tires, automobile parts, partition panels, wind mill blades, shoe soles, and several other technical applications.

Finally, we would like to conclude this chapter with an outlook presented by Liese [77]. Bamboo contributes to the existence of well over a billion people, mostly in the poor rural areas. Only because the wages of these peoples are low, bamboo is reasonable for others. Our obligation to improve the poor man's life quality will evenly affect the availability of bamboo to others. Reviewing the future of bamboo, four areas have been differentiated, emphasizing bamboo as a horticultural plant, as an essential community resource and a local source of income, but warning to overestimate the benefits from bamboo as an industrial resource. The bamboo we grow has to be utilized by and for the people with their environment. In the countries like ours, we are obliged to promote high-tech applications of bamboo fibers as a complementary activity to protect the natural environment and tradition of the people who grow with bamboos.

KEYWORDS

- **Bamboo**
- **Biocomposites**
- **Deformation Behavior**
- **Electron Microscopy**
- **Reinforced Plastics**

ACKNOWLEDGMENTS

The author thanks the Alexander von Humboldt (AvH) foundation for supporting his project on Natural Fiber Reinforcement of Polymers during his several short research visits to Germany. He further acknowledges Prof. Goerg H. Michler (Halle), Prof. Jean Marc Saiter (Rouen) and Prof. Sabu Thomas (Kottayam) for their unconditional moral supports and fruitful discussions on our green chemistry endeavors in Nepal. The author dedicates this chapter to Prof. F. J. Balta Calleja (CSIC, Madrid) on his 75th Birthday.

REFERENCES

1. Bledzki, A. K. and Gassan, J. Composites reinforced with cellulose based fibers. *Progress in Polymer Science*, **24**, 221-274 (1999).
2. Mohanty, A. K., Misra, M., and Drzal, L. T. *Natural Fibers, Biopolymers and Biocomposites*. CRC Press, Boca Raton, USA (2005).
3. Visakh, P. M. and Thomas, S. Preparation of bionanomaterials and their polymer nanocomposites from waste and biomass. *Waste Biomass Valor*, **1**, 121–134 (2010).
4. Klemm, D., Kramer, F., Moritz, S., Lindstroem, T., Ankerfors, M., Gray, D., and Dorris, A. Nanocelluloses: A new family of nature-based materials. *Angewandte Chemie, International Edition*, **50**, 5438-5466 (2011).
5. Siqueira, G., Tapin-Lingua, S., Bras, J., da Silva Perez, D., and Dufresne, A. (2011). Mechanical properties of natural rubber nanocomposites reinforced with cellulosic nanoparticles obtained from combined mechanical shearing, and enzymatic and acid hydrolysis of sisal fibers. *Cellulose*, **18**, 57-65.
6. Kalia, S., Kaith, B. S., and Kaur, I. Cellulose Fibers: Bio- and Nano-polymer Composites: *Green Chemistry and Technology*, Springer-Verlag, Berlin, Heidelberg, Germany (2011).
7. Jiang, L., Wolcott, M. P., and Zhang, J. Study of biodegradable polylactide/poly(butylenes adipate-co-terephthalate) blends. *Biomacromolecules*. **7**, 199-207 (2006).
8. Takagi, H. and Asano, A. Effects of processing conditions on flexural properties of cellulose nanofiber reinforced "green" composites. *Composites Part A*, **38**, 685-689 (2008).
9. Saiter, J. M., Dobircau, L., Saiah, R., Sreekumar, P. A., Galandon, A., Gattin, R., Leblanc, N., and Adhikari, R. Relaxation map of 100% green thermoplastic films: Glass transition and fragility. *Physica B*, **405**, 900-905 (2010).

10. Sreekumar, P. A., Gopalkrishnan, P., Leblanc, N., and Saiter, J. M. Effect of glycerol and short sisal fibers on the viscoelastic behavior of wheat flour based thermoplastic. *Composites A* **41**, 991-996 (2010).
11. Xu, Y., Wang, R.-H., Koutinas, A., and Webb, C. Microbial biodegradable plastics production from a wet-based biorefining strategy. *Progress in Biochemistry*, **45**, 153-163 (2010).
12. Adhikari, R., Bhandari, N. L., Causin, V., Le, H. H., Radusch, H. J., Michler, G. H., and Saiter, J. M. Fully Biodegradable Composites Based on Bamboo Flour as Filler. *Polymer Engineering and Science*, submitted (2011).
13. Xiao, Y. Modern Bamboo Structures (*Proceedings of the First International Conference on Modern Bamboo Structures* (ICBS-2007, Changsha, China, 28-30 October, 2007),Taylor & Francis, USA (2008).
14. Amada, S. and Lakes, R. S. Viscoelastic properties of bamboo. *Journal of Materials Science*, **32**, 2693-2697 (1997).
15. Scurlock, J. M. O., Dayton, D. C., and Hames, B. Bamboo: An overlooked biomass resource? *Biomass and Bioenergy*, **19**, 229-244 (2000).
16. Osorio, L., Trujillo, E., Van Vuure, A. W., and Verpoest, I. Morphological aspects and mechanical properties of single bamboo fibers and flexural characterization of bamboo/epoxy composites. *Journal of Reinforced Plastics and Composites*, **30**, 396–408 (2011).
17. Lima, H. C. Jr., Willrich, F. L., Barbosa, N. P., Rosa, M. A., and Cunha, B. S. Durability analysis of bamboo as concrete reinforcement. *Materials and Structures*, **41**, 981-989 (2008).
18. Sui, G. X., Yu, T. X., Kim, J. K., and Zhou, B. L. Mechanical behavior and failure modes, of aluminum/bamboo sandwich plates under quasi-static loading. *Journal of Materials Science*, **35**,1445-1452 (2000).
19. Ni, H., Li, X., Gao, H., and Nguyen, T. P. Nanoscale structural and mechanical characterization of bamboo-like polymer/silicon nanocomposite films. *Nanotechnology*, **16**, 1746–1753 (2005).
20. Yu, X. C., Sun, D. L., Sun, D. B., Xu, Z. H., and Li, X. S. Basic properties of wood ceramics made from bamboo powder and epoxy resin. *Wood Science & Technolology*, doi: 10.1007/s00226-010-0390-y (2011).
21. Toda, T., Okubo, K., Fujii, T., Hurutachi, H., Yamanaka, Y., and Yamamura, H. Development of rubber shoe sole containing bamboo fibers for frozen roads. *Conference paper presented during 16th International Conference on Composite Materials (ICCM-16)*, Kyoto, Japan (8-13 July, 2007).
22. Holmes, J. W., Brøndsted, P., Sørensen, B. F., Jiang, Z., Sun, Z., and Chen, X. Evaluation of a bamboo/epoxy composite as a potential material for hybrid wind turbine blades. *Paper presented at Global Wind Power*, Beijing (2008).
23. Grosser, D. and Liese, W. Verteilung der Leitbiindel und Zellarten in Sprogachsen verschiedener Bambusarten. *Holz als Roh- und Werkstoff*, **32**, 473-482 (1974).
24. Shin, F. G., Xian, X.-J., Zheng, W.-P., and Yipp, M. W. Analyses of the mechanical properties and microstructure of bamboo-epoxy composites. *Journal of Materials Science* **24**, 3483-3490 (1989).
25. Liese, W. The anatomy of bamboo culms. International Network for Bamboo and Rattan (INBAR), *Technical Report No*., **18**, ISBN 81-86247-26-2 (1998).
26. Schott, W. Bamboo under the microscope Online *Magazine*, www.powerfibers.com (Accessed Dec 10, 2011).
27. Rao, K. M. M. and Rao, K. M. Extraction and tensile properties of natural fibers: Vakka, date and bamboo. *Composite Structures*, **77**, 288–295 (2007).
28. Jain, S., Kumar, R., and Jindal, U. C. Mechanical behavior of bamboo and bamboo composite. *Journal of Materials Science* **27**, 4598-4604 (1992).
29. Wang, H., Chang, R., Sheng, K. C., Adl, M., and Qian, X. Q. Impact response of bamboo-plastic composites with the properties of bamboo and polyvinylchloride (PVC). *Journal of Bionic Engineering*, **5**, 28-33 (2008).
30. Chen, W., Yu, H., and Liu, Y. Preparation of millimeter-long cellulose I nanofibers with diameters of 30–80 nm from bamboo fibers. *Carbohydrate Polymers*, **86**, 453-461 (2011).

31. Zou, L., Jin, H., Lu, W. Y., and Li, X. Nanoscale structural and mechanical characterization of the cell wall of bamboo fibers. *Materials Science and Engineering: C*, **29**, 1375-1379 (2009).
32. Alvarez, V. A., Cyras, V. P., and Vazquez, A. Extraction of cellulose and preparation of nanocellulose from sisal fibers. *Cellulose*, **5**, 149–159 (2008).
33. Frone, A. N., Panaitescu, D. M., Donescu, D., Spataru, C. I., Radovici, C., Trusca, R., and Somoghi, R. Preparation and characterization of PVA composites with cellulose nanofibers obtained by ultrasonication. *BioResource*, **6**, 487-512 (2011).
34. Fuentes, C. A., Tran, L. Q. N., Dupont-Gillain, C., Vanderlinden, W., De Feyter, S., Van Vuure, A. W., and Verpoest, I. Wetting behaviour and surface properties of technical bamboo fibers. *Colloids and Surfaces A: Physicochemical and Engineering Aspects*, **380**, 89–99 (2011).
35. Ray, A. K., Das, S. K., Mondal, S., and Ramachandrarao, P. Microstructural characterization of bamboo. *Journal of Materials Science*, **39**, 1055-1060 (2004).
36. Chang, F., Lee, S. H., Toba, K., Nagatani, A., and Endo, T. Bamboo nanofiber preparation by HCW and grinding treatment and its application for nanocomposite. *Wood Science and Technology*, doi: 10.1007/s00226-011-0416-0 (2011).
37. Deshpande, A. Bhaskar Rao, P. and Lakshmana, Rao M. C. Extraction of bamboo fibers and their use as reinforcement in polymeric composites. *Journal of Applied Polymer Science*, **76**, 83–92 (2000).
38. Okubo, K., Fujii, T., and Yamamoto, Y. Development of bamboo-based polymer composites and their mechanical properties. *Composites: Part A*, **35**, 377-383 (2004).
39. Okubo, K., Fujii, T., and Thostenson, E. T. Multi-scale hybrid biocomposite: Processing and mechanical characterization of bamboo fiber reinforced PLA with micro fibrillated cellulose. *Composites: Part A*, **40**, 469–475 (2009).
40. Liu, D. Y., Yuan, X. W., Bhattacharyya, D., and Easteal, A. J. Characterization of solution cast cellulose nanofiber–reinforced poly(lactic acid). *eXPRESS Polymer Letters* **4**, 26-31 (2010).
41. Qua, E. H., Hornsby, P. R., Sharma, H. S. S., and Lyon, G. Preparation and characterization of cellulose nanofibers. *Journal of Materials Science*, **46**, 6029–6045 (2011).
42. Wang, Y., Leppanen, K., Andersson, S., Serimaa, R., Ren, H., and Fei, B. Studies on the nanostructure of the cell wall of bamboo using X-ray scattering. *Wood Science and Technology*, doi: 10.1007/s00226-011-0405-3 (2011).
43. He, J., Tang, Y., and Wang, S. Y. Differences in morphological characteristics of bamboo fibers and other natural cellulose fibers: studies on X-ray diffraction, Solid State 13C-CP/MAS NMR, and Second Derivative FTIR Spectroscopy Data. *Iranian Polymer Journal*, **16**, 807-818 (2007).
44. Bhat, I. H., Mustafa, M. T. B., Mohmod, A. L., and Abdul Khalil, H. P. S. Spectroscopic, thermal and anatomical characterization of cultivated bamboo. *BioResource*, **6**, 752-1763 (2011).
45. Kongkeaw, P., Nhuapeng, W., and Thamajaree, W. The effect of fiber length on tensile properties of epoxy resin composites reinforced by fibers of bamboo. *Journal of Microscopy Society of Thailand*, **4**, 46-48 (2011).
46. Kumar, V., Kushwaha, P. K., and Kumar, R. Impedance-spectroscopy analysis of oriented and mercerized bamboo fiber-reinforced epoxy composite. *Journal of Materials Science*, **46**, 3445–3451 (2011).
47. Ismail, H., Edyham, M. R., and Wirjosentono, B. Bamboo fiber filled natural rubber composites: the effects of filler loading and bonding agent. *Polymer Testing*, **21**, 139-144 (2002).
48. Ismail, H. The effects of filler loading and a silane coupling agent on the dynamic properties and swelling behaviour of bamboo filled natural rubber compounds. *Journal of Elastomers and Plastics*, **35**, 149-159 (2003).
49. Saxe, M. and Sorna Gowri, V. Studies on bamboo polymer composites with polyester amide polyol as interfacial agent. *Polymer Composites*, **24**, 428-436 (2003).
50. Lee, S. H., Ohkita, T., and Kitagawa, K. Eco-composite from poly(lactic acid) and bamboo fiber. *Holzforschung*, **58**, 529–536 (2004).
51. Tung, N. H., Yamamoto, H., Matsuoka, T., and Fuji, T. Effect of surface treatment on interfacial strength between bamboo fiber and PP resin. *JSME International Journal, Series A*, **47**, 561-565 (2004).

52. Lee, S. H. and Wang, S. Biodegradable polymers/bamboo fiber biocomposite with bio-based coupling agent. *Composites: Part A* **37**, 80–91 (2006).
53. Han, G., Lei, Y., Wu, Q., Kojima, Y., and Suzuki, S. Bamboo–fiber filled high density polyethylene composites: effect of coupling treatment and nanoclay. *Journal of Polymers and Environment*, **16**, 123-130 (2008).
54. Liu, H., Chen, F., Liu, B., Estep, G., and Zhang, J. Super toughened poly(lactic acid) ternary blends by simultaneous dynamic vulcanization and interfacial compatibilization. *Macromolecules*, **43**, 6058-6066 (2010b).
55. Mi, Y., Chen, X., and Guo, Q. Bamboo fiber-reinforced polypropylene composites: Crystallization and interfacial morphology. *Journal of Applied Polymer Science*, **64**, 1267–1273 (1997).
56. Chen, X., Guo, Q., and Mi, Y. Bamboo fiber reijnforced polypropylene composites: A study of the mechanical properties. *Journal of Applied Polymer Science* **69**, 891-1899 (1998).
57. Trujillo, E., Osorio, L., Van Vuure, A. W., Ivens, J., and Verpoest, I. Characterization of polymer composite materials based on bamboo fibers. *Paper ID: 344-ECCM14, 14th European Conference on Composite Materials*, Budapest, Hungary (June 7–10, 2010).
58. Ying-Chen, Z., Hong-Yan, W., and Yi-Ping, Q. Morphology and properties of hybrid composites based on polypropylene/polylactic acid blend and bamboo fiber. *Bioresource Technology*, **101**, 7944-7950 (2010).
59. Xu, Y., Lee, S. Y., and Wu, Q. Creep analysis of bamboo high-density polyethylene composites: Effect of interfacial treatment and fiber loading level. *Polymer Composites*, **32**, 692–699 (2011).
60. Siro, I. and Plackett, D. Microfibrillated cellulose and new nanocomposite materials: A review. *Cellulose*, **17**, 459-494 (2010).
61. Kori, Y., Kitagawa, K., and Hamada, H. Crystallization behavior and viscoelasticity of bamboo-fiber composites. *Journal of Applied Polymer Science*, **98**, 603–612 (2005).
62. Henning, S., Adhikari, R., Michler,, G. H., Baltá Calleja, F. J., and Karger-Kocsis, J. Micromechanical mechanisms for toughness enhancement in β-modified polypropylene. *Macromolecular Symposia*, **214**, 157–172 (2004).
63. Rajulu, A. V., Narasimha Chary, K., Ramachandra Reddy, G., and Meng, Y. Z. Void content, density and weight reduction studies on short bamboo fiber–epoxy composites. *Journal of Reinforced Plastics and Composites*, **23**, 127-130 (2004).
64. Kushwaha, P. K. and Kumar, R. Effect of silanes on mechanical properties of bamboo fiber-epoxy composites. *Journal of Reinforced Plastics and Composites*, **29**, 718-724 (2010).
65. Raghavendra Rao, H., Rajulu, A. V., Ramachandra Reddy, G., and Hemachandra Reddy, K. Flexural and compressive properties of bamboo and glass fiber-reinforced epoxy hybrid composites. *Journal of Reinforced Plastics and Composites*, **29**, 1446-1450 (2010).
66. Kushwaha, P. K. and Kumar, R. Enhanced mechanical strength of BFRP composite using modified bamboos. *Journal of Reinforced Plastics and Composites*, **28**, 2851-2859 (2009).
67. Das, M. and Chakraborty, D. The effect of alkalization and fiber loading on the mechanical properties of bamboo fiber composites, part 1: Polyester resin matrix. *Journal of Applied Polymer Science*, **112**, 489-495 (2009).
68. Wong, K. J., Zahi, S., Low, K. O., and Lim, C. C. Fracture characterization of short bamboo fiber reinforced polyester composites. *Materials and Design* **31**, 4147-4154 (2010).
69. Ratna Prasad, A.V. and Mohana Rao, K. Mechanical properties of natural fiber reinforced polyester composites: Jowar, sisal and bamboo. *Materials and Design* **32**, 4658-4663 (2011).
70. Satyanarayana, K. G., Arizaga, G. G. C., and Wypych, F Biodegradable composites based on lignocellulosic fibers an overview. *Progress in Polymer Science*, **34**, 982-1021 (2009).
71. Jiang, L., Chen, F., Qian, J., Huang, J., Wolcott, M., Liu, L., and Zhang, J. Reinforcing and toughening effects of bamboo pulp fiber on poly(3-hydroxybutyrate-co-3-hydroxyvalerate) fiber composites. *Industrial and. Engineering Chemistry Research*, **49**, 572–577 (2010).
72. Okubo, K., Fujii, T., and Yamamoto, Y. Improvement of interfacial adhesion in bamboo polymer composite enhanced with microfibrillated cellulose. *JSME International Journal, Series A*, **48**, 199-204 (2005).

73. Kitagawa, K., Ishiaku, U. S., Mizoguchi, M., and Hamada, H. Bamboo-based ecocomposites and their potential applications. In *Natural Fibers, Biopolymers, and Biocomposites*. A. K. Mohanty, M. Misra, and L. T. Drzal (Eds.). CRC Press, Boka Raton, USA (2005).
74. Rychter, P., Kawalec, M., Sobota, M., Kurcok, P., and Kowalczuk, M. Study of aliphatic-aromatic copolyester degradation in sandy soil and its cotoxicological impact. *Biomacromolecules*, **11**, 839–847 (2010).
75. Kleeberg, I., Hetz, C., Michael-Kroppenstedt, R., Müller, R. J., and Deckwer, W. D. Biodegradation of aliphatic-aromatic copolyesters by thermomonospora fusca and other thermophilic compost isolates. *Applied Environmental Microbiology*, **64**, 1731-1736 (1998).
76. Witt, U., Einig, T., Yamamoto, M., Kleeberg, I., Deckwer, W. D., Müller, R. J. Biodegradation of aliphatic-aromatic copolyesters: Evaluation of the final biodegradability and ecotoxicological impact of degradation intermediates. *Chemosphere*, **44**, 289–299 (2001).
77. Liese, W. *Bamboo: Past-Present-Future*. American Bamboo Society, Newsletter, vol. 20 (February, 1999).

14 Green Methods to Synthesize and Recycle Materials: A Promise to the Future

P. Deepalekshmi and Sabu Thomas

CONTENTS

14.1 INTRODUCTION

The increase in the energy demand forces human kind to find out new alternatives to existing sources. Engineering materials applicable in medical, industrial, and technological field have been developed to great extents during the past decade. Several physical as well as chemical reactions are practiced in order to synthesize new materials for various applications. However, the disadvantages of these methods are not simple and cannot be discarded easily. The long reaction time needed for the reactions, use of harmful and toxic catalysts, use of chlorinated, and often high boiling point solvents and so on are only a few of them. Most of the preparation methods involve the use of toxic solvents, release of unwanted pollutants to the atmosphere, and complicated steps. In order to alleviate the impact of these problems, cleaner and safer methods were developed. Earlier scientists were concentrating on the yield, easiness, and fastness of reactions, but now the world is looking for environmental friendly methods as well. Due to this fact, apart from the lower fabrication costs and improved material properties, environmental concern is also anticipated for the synthesis of new materials. All the processes eliminating the use of harmful chemicals and thus not imposing any problem to living cells are regarded as green methods. Such eco-friendly methods include the use of ionic liquids, microwave irradiation, solid phase syntheses, and so on. The influence of ionic liquids and microwave assistant green methods in organic synthesis has been well explained by Palou et al. [1]. An example of green chemical method is seen in rail transportation system. This mode of transportation emits a large amount of green house gases to the atmosphere, and this can be alleviated by establishing a green method. By using aluminum alloy, composite material, and sandwich panel for rail car bodies and interiors and thereby adopting a lightweight material technology, emission of gases can be controlled to some extent. The development of carbon fiber reinforced plastic is yet another way to decrease the weight of the car bodies and thus to reduce the fuel consumption.

As a consequence of population, energy crisis, and technological revolutions tremendous amount of waste materials are also released to the atmosphere, eventually creating a lot of health hazards to all living beings. Nanotechnology donated a large number of industrial materials to the world, but unless both its synthesis and usage are carefully monitored, it can become a threat to life. In this context, the waste disposal management has high significance. The dumping of wastes is not a solution to escape from its bad impact. Due to this fact, most of the industries are based on the recovery, reuse, and recycling procedures. Of the several ways of waste disposal, recycling is the most important one, but the high cost involved in it and the environmental problems make it less practical. Yet quiet a lot recycling processes are running all over the world. Recycling of materials should be economically and scientifically acceptable by a majority of the society and must be environmentally friendly. But unfortunately in most cases, the disposal methods such as incineration and landfill are found to be more practical and environmentally friendly than recycle. The disposal of waste materials based on the nature is the most wise and practically used method and thus green methods have much significance.

Green methods are flexible, adaptable, and environmentally friendly and would help to meet all requirements of the current situation. The easiness and simplicity of the method helps to develop the product of the reaction within short periods of time avoiding the prolonged conventional procedure. The important advantages of these methods are summarized in Figure 1.

Nanomaterials have several applications in various fields of life like medicine, agriculture, industry, energy, and so on. Several chemical methods are reported to be successful for the synthesis of nanoparticles as well as their composites. For example, the synthesis of nanoparticles can be possible by reduction reactions. Apart from the nanoparticles, polymers are another promising class of materials. By combining the possibilities of these two, another new class is also emerged, Polymer nanocomposites which is yet another area of study. Even though all these compounds can be prepared extensively, there mode of preparation and recycling has some significance. Due to the increased importance of these promising materials, we discuss some of the greener ways to synthesize and recycle them in this chapter. Main focus is given to metal nanoparticles having various applications especially in the medical field and also to the polymers used in almost all areas of life. Polymer wastes are the most recycled wastes among others, but greatly by chemical methods. Thus the greener ways to recycle polymers is really important and it is also a topic of study of this chapter.

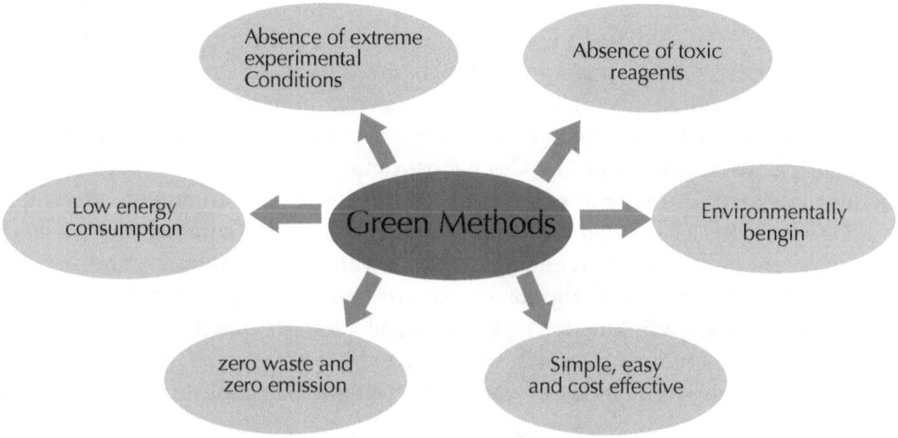

FIGURE 1 Advantages of green methods.

14.2 GREENER WAYS OF SYNTHESIS

14.2.1 Synthesis of Metal Particles

Because of the diverse applications of metallic nanoparticles in areas such as electronics, cosmetics, coatings, packaging, and biotechnology, their synthesis is attaining great attention. The two approaches for synthesis of metallic nanoparticles include the bottom up or self assembly process and the top down process. Bottom up method gives nanoparticles with lesser defects and more homogeneity [2]. Even though the synthesis can be done by physical as well as chemical means, these involve the use of toxic solvents, generation of hazardos by products, and high energy consumption. So, exploring biological green routes is really essential. Green way of synthesis of materials is cost effective, environment friendly; easily scaled up for large scale synthesis and in this method there is no need to use high pressure, energy, temperature, and toxic chemicals. Silver nanoparticles are synthesized from plant extracts by Bunghez et al [3]. They reacted plant extract with aqueous $AgNO_3$ solution and confirmed the bioreduction of Ag^+ ions. The antibacterial activity of silver nanoparticles can be utilized by this method [3]. Another study based on polyalthialongifolia leaf extract was reported to be depending on the temperature conditions. The particles synthesized at high temperature were smaller and have better antibacterial activity. The particles are toxic to both gram positive and gram negative bacteria. [4] The extracellular synthesis of silver particles can be done by *Bacillus Subtillus*. This bacterial reduction took very short time to complete and the synthesized nanoparticles can be used in biological sensors, drug, and gene delivery and antimicrobial protection [5]. They found the produced nanoparticles to be most effective against clinically isolated human pathogens. Silver nanoparticles have many medicinal applications like in wound dressing, textile fabrics for burnt patients, and also in surgical masks.

Silver nanoparticles can also be prepared by reducing Ag^+ ions using polyethylene glycol. Due to the reusability and recyclability of this polymer, the method is very appreciable for the nanoparticle synthesis. Using these particles, Yan et al [6] coupled aldehydes, alkynes, and amines in one pot and produced propargylamines in high yields [6]. By avoiding the conventionally used toxic trioctylphosphine, Beri et al. [7] synthesized high quality ZnSe quantum dots using zinc acetate and oleic acid [7]. This invention has much significance as quantum dots especially ZnSe offers wide applications such as in diode laser structures, green–blue light emission and in solar cells. Cauliflower like CuI nanostructures useful as catalyst in coupling reactions and also for the removal of heavy metal ion like Cadmium are developed by Jiang et al [8]. They used the mild, inexpensive, water soluble, and nontoxic reducing agent ampicillin as reducing agent as well as morphology stabilizer. The high specific surface area of the nanostructure makes it applicable in adsorption of materials. The growth kinetics of Ag nanoparticles is studied by [9], synthesizing the particles by reducing $AgNO_3$ using the extract of neem (*Azadirachtaindica*) leaves. The synthesize methods of various metal nanoparticles are summarized in Table 1.

TABLE 1 Green synthesize routes for metal nanoparticles.

Nanoparticle	Synthesis Route	Advantage	Application
$CuInS_2$ $CuInSe_2$ [10]	Microwave heating of Cu, In and Se/S precursors in water in the presence of mercapto-aceticacid	Elimination of organic solvent	Photovoltaic solar cell
Ag [11]	Reducton of Ag+ using medicinal plant, *Meliaazedarach*	Tannic acid in plant extract initiates the synthesis of Ag and stabilizes it.	Superior cytotoxicity
Ag [12]	Reduction of $AgNO_3$ by glucose in the presence of gelatin (stabilizer) and NaOH(accelerator)	Use of natural polymers. Method is rapid and easy. Gelatin prevents aggregation.	In chemical and biological industries for getting individual nanoparticles.
Ag [13]	using extract of a fennel, *Foeniculumvulgare*	Fennel has medicinal value and is a carminative	Antibacterial activity against two human pathogenic bacteria *S. aureus* and *E. coli i* when combined with antibiotics
Au [14]	Reduction of chloro auric solution using cell-free filtrate of fungus, *Sclerotiumrolfsii.*	Rapid synthesis of isotropic and anisotropic particles	Applicable in cancer hyperthermia

TABLE 1 *(Continued)*

Nanoparticle	Synthesis Route	Advantage	Application
Ag colloid [5]	PEG acts both as reducing agent and stabilizer in water solvent.	PEG prevents aggregation of AG particles	Antibacterial against Gram-negative *E. coli* and *P. aeruginosa*, and Gram-positive *S. aureus*, and fungi *C. albicans*
Ag [16]	Reduction of $AgNO_3$ by carboxy methyl cellulose at suitable pH and temperature	Ag nanoparticles with concentration as high as 1000 rpm is prepared	Antimicrobial agents for textiles.
Ag [17]	Reducing $AgNO_3$ using sulfated polysaccharide isolated from marine red algae (Porphyravietnamensis)	Size controlled synthesis of Ag particles and wide range of stability from pH 210	Selective antibacterial activity against Gram negative bacteria
MoS_2, Ag_2S, CdS [18]	composite-surfactants-aided-solvothermal process	Controllable particle size and shape by varying the experimental conditions	Useful in fabrication of nanoscale materials
Au [19]	Using naturally occurring red ginseng root	One step process in which phytochemicals act as coatings which prevent agglomeration and excellent in vitro stability.	Applicable in optoelectronic and biomedical fields due to the non toxicity.
Ag [20]	Hydrolysis of carboxy methyl cellulose by microwave irradiation and reducing Ag+ ions by the hydrolyzate	Homogeneous heating by microwave and so uniform and stable particles, which can be stored at room temperature for long period.	Medical field.

14.2.2 Synthesis of Composites

Polymers are widely used as stabilizers during the synthesis of metal nanoparticles due to their ability to control particle growth and to prevent the oxidation of particles. Among several polymers, polyethylene glycol, a water soluble polymer is widely used for the synthesis of Ag particles. Using $AgNO_3$ as the precursor for the synthesis, PEG/Chitosan/Ag nanocomposite is reported by Ahmad et al [21]. The composite preparation included two subsequent steps consisting of synthesizing the Ag nanoparticles in the first step followed by the preparation of the composite. The stirring time is found to affect the formation of smaller Ag particles. The increased stirring time during the process causes better compatibility in the nanocomposites. Cellulose/Ag nanocomposite having prolonged antibacterial properties are produced using bacterial cellulose (BC), a special kind of cellulose synthesized from bacteria [22]. When compared to the natural cellulose, BC has three dimensional nanofiber network structures and does not have lignin and hemicelluloses components in it. This product has unique properties such as high purity, porosity and textile strength, increased degree of polymerization, and good biocompatibility. The hydroxyl groups on BC act both as reducing and stabilizing agents for the synthesis of nanoparticle. The prepared BC/Ag nanocomposite by hydrothermal process can release Ag particles for a long term. This makes the composite useful in various antibacterial applications including medical devices, health care; self-sterilizing textiles, water purification, and food packages. Graphene nanosheets/Zinc oxide nano composites useful in making super capacitor materials are also prepared by a green route [23]. The nanosheets are prepared using glucose as the reducing agent and graphene oxide as the precursor and thereafter the ZnO nanoparticles are grown on the sheets. These spherical particles help the sheets to separate thus preventing the agglomeration of graphene. This results in the improved capacitance of composite electrode and conductivity. A biomolecule assisted hydrothermal method is reported for producing Ag/ZnO metal semiconductor nanocomposites [24]. In this organic combination of green chemistry and functional materials, bovine serum albumin is used both as a shape controller and as a reducing agent for Ag^+ ions. The advantages of this method include simple experimental procedure, lower reaction time and temperature, and obtained uniform morphology for samples. Since Ag nanoparticles on the ZnO act as electron sinks, the photocatalytic activity of ZnO is enhanced by this method. Photo catalysts applicable in both environmental purification and energy production processes can be made by following this method.

14.2.3 Synthesis of Polymer Based Systems

Carbiondioxide is proved to be an eco-friendly solvent for the synthesis of polymers and also for several polymer based applications. Polymers such as fluoropolymers, dispersion polymers, and step growth polymers are synthesized using carbon dioxide [25]. Requirements of any additional synthesis methods as well as the easy preparation of supercritical and liquid CO_2 are the major advantages of carbon dioxide based methods. Both soluble and insoluble polymers are prepared by homogeneous solution polymerization and heterogeneous chain growth methods. The preparation of fluoropolymers by this method eliminates the use of chlorofluorocarbons and thus saves ozone layer. Since many monomers and initiators are soluble in CO_2 medium and their polymer is insoluble, dispersion polymerization can be effectively done using this solvent. Polymers act as surfactants in Carbon dioxide making it applicable in the extraction, coating, and cleaning industries. The efficiency of carbon dioxide in polymer manufacture was explored by Wood et al [26]. The importance of supercritical carbon dioxide during the processing of fland cleaning and porous materials is explained well. They reported carbon dioxide as a reagent which is environmentally friendly, nontoxic, nonflclepor, inexpensive, and readily available in high purity from a number of sources. Safety in handling and easiness in the product isolation from its gaseous state are considered to be additional advantages of this material. Macroporous polystyrene particles synthesis by suspension polymerization in water and super critical CO_2 using polyvinyl alcohol (PVA) as stabilizer, divinylbenzene (DVB) as cross-linker, and azobisisobutyronitrile (AIBN) as initiator is reported [27]. The solvent properties can be varied by changing the pressure and density. Macroporous polymers applicable in various fields like ion exchange resins, chromatographic separation media, solid-supported reagents and so on are generally prepared by polymerisation reaction in the presence of suitable porogens. These organic solvents (porogens) cannot be removed completely from the polymer after the reaction, thus making these materials unsuitable in applications where high purity is needed. This problem can be solved by the use of supercritical CO_2 when used as a porogen. By controlling the pressure, the pore size, pore size distribution, surface area, and the average diameter of the particles can be regulated. The surface area, diameter and pore size of the particles are found to be increasing with increase in pressure. The greatest advantage of this method is that by simply depressurizing, the porogen can be completely removed from the medium.

Green composites are prepared based on epoxidized soybean oil, in which the fibers and nanoclay provided additional reinforcement [28]. Natural fibers like flax and clay are notable materials to provide reinforcement to polymer systems because of their strength and cost effectiveness. The flexural strength and tensile modulus of the obtained green composites are found to increase with the fiber loading but upto an optimum value, and thereafter the strength decreases. The organoclay nanocomposites are good alternatives to petrochemical polymers. Polyethylene glycol is a novel multifunctional polymer and its synthesis is really important in chemistry. Due to its multivalent sites and high drug carrying capacity, PEG dendrimers are widely used in drug delivery systems. Puskas et al. [29] made these dendrimers by following a green method of enzymatic catalysis. The reaction includes enzyme catalyzed transesterification reaction and Michael addition. Four functional dendrimers having multiple targeting, therapeutic, imaging, and diagnostic agents were reported from this work. The use of toxic solvents is the major problem that makes the production of polymers non eco-friendly and the replacement of such solvents by environmental friendly materials is the topic of study for several years. Various polymethacrilamides from methacryilic acid and amine are manufactured using microwave assisted solvent free method [30]. The microwave assisted synthesis took place in one step and it does not require any further activation. The polymer is formed by this easy single step method in 60% yield. The development of a biocompatible polymer from a bio renewable source is established by Okuda. He reported the synthesis of isotactic poly lactic acid based on tin initiators and found that it is stereoselective. They tried to find out the structure stereoselectivity relationship by using the metal complexes of Lewis acids. These metal complexes participate in the reaction mechanism and cause hetero selectivity [31].

Excellent weather resistance, resistance to alkali and hot water as well as high crystallinity makes polyvinyl alcohol to be used as basic materials for high modulus/high strength fibers. For the manufacture of PVA, Poly (vinyl pivalate) (PVPi) is found to be the best precursor material and its synthesis is a matter of importance due to their biodegradable nature. This PVPi is obtained by dispersion polymerization in supercritical carbon dioxide medium under high pressure [32] and the further hydrolysis can lead to the formation of syndiotactic PVA. They also studied about the influence of pressure, reaction time, initiator concentration, and monomer loadings on the PVA synthesis. As biodegradable and flame retardant materials, phosphorous containing polymers too have significant applications. To synthesis these polymers, interfacial polycondensation was used. Iliescu et al. [33] introduced a new green method for the successful production of these materials by a solvent and catalyst free gas–liquid and a solid–liquid interfacial polycondensation reaction. In the gas liquid polycondensation method, only water is present other than the reagents and hydrolyzing monomers can be used. Since the polymer is insoluble in water, the usually present complicated purification step can be avoided. The solid liquid method suppresses side reactions. The catalyst free conditions and green solvent together cause reduction in reaction times and enhancements in yields. When the solid liquid as well as the gas liquid method is compared, high molecular weight polymers are formed by the former process [33].

14.3 GREEN WAYS OF RECYCLING

Recycling of materials by conventional methods often involve the usage of large amount of energy. Also recycling factories release pollutants to atmosphere causing serious health hazards. When polymers such as rubbers are incinerated it liberates much heat energy. A lot of care should be given in this case as it can lead to serious issues like explosions. Another problematic matter is the use of acids and other chemicals during the recycling reactions. Biological recycling using bacteria is a solution to these kinds of issues but all sorts of materials cannot be recycled by this way. Also the maintenance of experimental conditions with bio organisms needs special care and equipments. Green recycling offers a cleaner and safer environment by eliminating all the bad impacts of conventional physical or chemical recycling processes. The need of green recycling over the conventional polluted recycling is clear from the Figure 2. The recycling methods for a few important chemical substances and polymers that we use in our daily lives are discussed in this section.

FIGURE 2 Advantages of green recycling.

14.4 RECYCLING OF COMPOUNDS

Tetrahydropyran ring system is very important for organic syntheses. Organic chemists often need to prepare this, since it is the backbone of several important carbohydrates, antibiotics, and natural products. But many of the methods used for the synthesis of this ring compound produces harmful toxic wastes. The use of chlorinated solvents is not welcomed due to the environmental problems and high waste disposal cost. Yadav et al [34] showed the efficiency of green ionic liquids as solvents for various organic reactions [34]. The tunable polarity, high thermal stability, and immiscibility with a number of organic solvents, negligible vapor pressure, and recyclability make these solvents capable to replace the conventional toxic solvents. The advantage of ionic liquids not only lies in the separation of product catalyst mixture but also in the reusability of it. The yield of the product is not found to be decreasing when the recycled ionic liquid is used for the reaction. The mild reaction conditions, simplicity in operation, improved yields and reaction rates, cleaner reaction profiles, and recyclability of ionic liquids make the process a simple, convenient, and eco-friendly compared to organic solvents. The use of ionic liquids in controlling the organic pollutants in waste water, waste gas, solid wastes, and in contaminated soil is studied by Ma and Hong [35]. They focused mainly on some of the common pollutants in nature like phenolic compounds, polycyclic aromatic hydrocarbon (PAHs), dyes from wastewater, chlorophenothane (DDT), and dieldrin and established the importance of ionic liquids in recycling them. Even though the ionic liquids itself is not completely devoid of toxicity and environmental problems, recycling of these solvents by simple as well as cost effective means such as distillation seems to be a good solution.

Dumping, open burning, or open detonation procedures are some of the traditional methods of disposal of old ammunition used in defense. Air defiance ammunition has propulsion part as well as warhead part. The propulsion part contains a small amount of primary explosive and larger quantities of the propellant whereas the warhead part consists of a secondary explosive such as TNT or Hexal and small amounts of pyrotechnics, and primary explosive. Since the propellants are metal free and contain nitrocellulose, its recycling is very easy. The most difficult component to be recycled is hexal, which behaves as a hard material under high pressure. Even though there are several methods for the recycling of hexal, all the processes require high investment costs and causes pollution. The traditional open detonation method causes noise pollution and open burning produces toxic nitric oxides. During the recycling of warhead also, hexal cannot be recycled. Brogle et al. [36] reported an effective method of extracting hexal. This is done by drilling a hole in the hexal, and then applying a high pressure water jet at a pressure of several hundred bars to remove hexal. This can be reused as a basic component for the formation of new hexal. The water used in this process can also be purified [36], thus making it environmentally benign. Lead is a very important constituent of batteries, solderings, and dielectric materials. Even though lead containing materials are now replaced by non lead components due to its toxicity, the disposal of this element is a matter of great importance. Lead containing glass is very essential in radiation technologies as it prevents the passage of radioactive radiations. Since lead atoms are encapsulated inside the silicon and oxygen atoms in the glass structure, their recycling is really difficult unless a very high temperature and pressure is applied. So most spent lead glass has been land find after some detoxification and stabilization processes, for their disposal. Sasai et al. [37] introduced an eco-friendly simple mechano chemical method for the recycling of lead glass. The method involves grinding and mixing of materials with high mechanical energy so that the particles having more surface area are generated. This helps to break the bonding between the atoms and to provide an easy path for recycling. They investigated the extraction of lead from the lead glass powder by wet ball milling treatment with Na_2EDTA chelate reagent, then separating the sodium-EDTA and lead-EDTA solutions and finally the eco-friendly recycling method in which most of the elements are recycled [37].

14.5 RECYCLING OF POLYMERS

The light weight, flexibility, and versatility of polymers make them suitable for a large number of applications and uses. Their optimum design with functional solutions, economically affordable nature, and extreme durability replaces many other materials used so far. The most important property of this systems is their possibility of recovery and reuse and thus again contributing to sustainable development and environmental protection. Plastic materials have application in almost all aspects of modern society but making their waste management a challenge both from the economical and ecological point of view. Due to the high cost of recycling, large amounts of post consumer plastics have been wasted by being deposited in landfills nowadays. This emphasizes on the necessity of recycling. Several recycling methods for plastic products with doubtful ecological benefits have led to low value products in the past which were difficult to market. The major recycling methods of polymers include Chemical, Mechanical, Thermal and combinations of these. Biological methods are also possible in certain cases, but the maintenance of proper experimental conditions prevents this method from their wide use. Among those recycling technologies, mechanical recycling generally yields a high environmental benefit [38]. The effective plastic waste management system can reduce CO_2 emissions and can save fossil fuels.

In the waste generation mainly in the production of greenhouse gases and hazardous substance emissions, automobile industry stands first among all the other existing industries. Depending on the various components used for making the automobiles, the recycling techniques vary to a wide extent. The different components are separated before recycling as ferrous, nonferrous, plastics and rubbers and should be processed separately. The recycling of all types of components without causing any side effects is a matter of serious concern in the automobile industry. Metal Plated Plastics (MPP) is used in automobile parts because of its high performance and easiness to give any shape by molding it. The MPPs are made of Acrylonitrile butadiene styrene copolymer system or ABS/polycarbonate and Copper, Nickel, and Chromium are generally used for the metal parts. These MPPs are very essential for the automobile industry, but their disposal is a very big problem. Its recycling has to be considered seriously not only because of the energy resources point of view, but the disposal of these materials can cause serious environmental problems. The chemical as well as electrochemical processes used currently for the recycling of MPPs have several limitations. Due to the complicated processing steps involved and the destruction of plastics after recycling, this method cannot be recommended. Moreover it utilizes much chemical reagents and causes the formation of toxic gases and wastes during recycling thereby making it non eco-friendly. Xue et al. [39] developed an eco-friendly magnetic separation method for the recycling of these MPPS. In this method, firstly the MPPs are crushed using a self-designed hammer crusher and then coatings are removed from it by magnetic separation. The process is repeated until it is completely liberated. The efficient recovery of resources and prevention of the secondary pollution caused by reducing gas and acid liquid generated in the acid washing process are the advantages of this process. Also no waste water generation happens during the procedure. The liberated plastics are reutilized by granulating and the coatings as medium alloy.

Bio-based polymers based on natural products are gaining importance as they can be easily bio-assimilated by hydro-biodegradation. But the need of chemical modification for these kinds of polymers makes them less suitable from wide usage. Plastics have satisfactory technological properties but it must also be modified to make it oxo-biodegradable. The energy required for the recycling and recovery processes of these materials is really high. Because of manufacture and post consumer disposal, polyolefin's are found to be greener materials than bio-based polymers. The incineration of polyolefin's followed by heat recovery or their mechanical recycling utilizes the whole energy content of the plastics and this is greater than the energy used for recovery or recycling. Peroxidation can change polyolefin's to be biodegradable and the rate of peroxidation can be controlled easily by using environmentally sensitive antioxidants and light stabilizers [40]. Poly Lactic Acid (PLA) is a widely studied biocompatible and biodegradable polymer for several applications especially medical. Because of its wide range of applicability the production of PLA has increased to higher extents and this consumes energy, which is derived from renewable sources. To dispose the wastes efficiently the biodegradation of this polymer should be carried out on an industrial basis. More clearly, as PLA becomes more widely used and need disposal on a large scale it is reliable not to depend completely on its biodegradation either in the environment or in composting facilities, but suitable technologies must be practiced. Thus the polymer can be made more sustainable [41]. Since there is a possibility for polymer degradation during milling and extrusion, mechanical processing is inappropriate for PLA recycling. So even the process is applied, during each step the product obtained will have lower quality than the starting material making the market value of the product decreased. Other useful chemical methods like thermal degradation and hydrolysis are possible for PLA recycling, producing lactide by the former, and lactic acid by the latter. A green method using transesterification of PLA with alcohols to lactate esters is reported. They observed that the post consumer polyethylene terephthalate (PET) waste can influence the recycling of PLA and can decrease its market value. By using various catalysts glycolysis of PET is done and zinc acetate is identified to be the most powerful catalyst. Using this catalyst, the transesterification of PLA is done and the effect of processing conditions on the selective depolymerisation of PLA is investigated from a mixture of PLA and PET using the solvents, ethanol and methanol. The reactions cause PET to separate as a solid material and can be removed by filtration. The yield is improved when methanol is used as the solvent and in both cases PET remained unreacted. This type of recycling procedure can be applicable for a copolymer of PLA and PET as well.

Sato et al. [42] explored environmental friendly chemical recycling processes involving the decomposition of polymers to corresponding monomers using sub and super critical water as the reagents. The advantage of this method is the absence of catalysts during the chemical depolymerization. They investigated the decomposition of two polymers of PET and PEN (poly(ethylene 2,6-naphthalene dicarboxylate). The polymers decay in to their corresponding monomers at high temperature in water. Super critical water can hydrolyze the polymers and ethylene glycol is formed as a by-product in both cases. When the temperature is increased from 523 to 623K, the yields of monomers increase up to 80% whereas the amount of ethylene glycol decreases. It is observed from the stability studies with and without acids that, the lower yield of ethylene glycol in the depolymerization process was due to its dehydroxylation to aldehyde and polymerization to diethylene glycols, which is catalyzed by the protons of corresponding monomers dicarboxylic acids. Super critical technology is also applied to recycle the carbon reinforced plastics [43]. Due to low thermal expansion, high fatigue resistance, high mechanical strength, and good resistance to corrosion, carbon reinforced composites are found to be useful in aerospace, nuclear sector, and industries.

As mentioned under the recycling methods for compounds, due to environmental feasibility, ionic liquids are useful in recycling certain polymers as well [35]. They reported the recycling of polyamides from solid wastes by following treatment with ionic liquids. Plastics from solid wastes are recovered using the liquid,1-hexylpyridinium bromide [44]. The methanolysis of Poly Carbonate to obtain its starting monomers at nearly 100% yield is carried out using an ionic liquid [Bmim][Ac] [45]. This process evaded the use of traditional acid and base catalysts. The authors also studied about the kinetics of the reaction and reported that it follows first order. All these greener ways of recycling can eliminate most of the difficulties such as the equipment corrosion, pollutant emission, tedious work up procedures, and so on and thus opens its way up to the most modern polymer technology.

14.6 CONCLUSION

Nanoparticles and polymer composites are two important areas of study because of their wide range of applicability. Metal nanoparticles are abundantly synthesized because of its vast applicability in various fields such as industry, medicine, and technology. Upon considering the applications, polymers also impart unavoidable contributions. More studies are necessary on their synthesis and recycling methods as both are creating a lot of environmental hazards. Most of the recent methods consume a lot of energy and produce tremendous amount of waste materials. The use of toxic chemicals and emission of lump sum of pollutants are certainly a threat to mankind. In this regard, green methods have most significance. This chapter is a general discussion about some of the recent eco-friendly methods used for synthesizing as well as recycling these materials. By following such ways of processing, environment can be protected for a better future.

KEYWORDS

- **Ammunition**
- **Depolymerization Process**
- **Metal Plated Plastics**
- **Polymer Nanocomposites**
- **Recycling**

REFERENCES

1. Palou, R. M. Ionic Liquid and Microwave-Assisted Organic Synthesis:A "Green" and Synergic Couple. *J. Mex. Chem. Soc.*, **51**(4), 252264 (2007).
2. Thakkar, K. N., Mhatre, S. S., and Parikh, R. Y. Biological synthesis of metallic nanoparticles. *Nanomedicine: Nanotechnology, Biology, and Medicine*, **6**, 257–262 (2010).
3. Bunghez, I. R., Ghiurea, M., Faraon, V., and Ion, R. M. Green synthesis of silver nanoparticles obtained from plant extracts and their antimicrobial activities. *Journal of Optoelectronics and advanced materials*, **13**(7), 870–873 (2011).
4. Kaviya, S., Santhanalakshmi, J., and Viswanathan, B. Green Synthesis of Silver Nanoparticles Using Polyalthialongifolia Leaf Extract along with D-Sorbitol: Study of Antibacterial Activity. *Journal of Nanotechnology, Article ID 152970*, 5 (2011).
5. Kannan, N. and Subbalaxmi, S. (2011).Green synthesis of silver nanoparticles using Bacillus Subtillus IA751 and its Antimicrobial Activity. *Research Journal of Nanoscience and Nanotechnology*, **1**(2), 8794.
6. Yan, W., Wang, R., Xu, Z., Xu, J., Lin, L., Shen, Z., and Zhou, Y. A novel, practical and green synthesis of Ag nanoparticles catalyst and its application in three-component coupling of aldehyde, alkyne, and amine. *Journal of Molecular Catalysis A: Chemical*, **255**, 81–85 (2006).
7. Beri, R. K., More, P., Bharate, B. G., and Khanna, P. K. Band-gap engineering of ZnSe quantum dots *via* a non-TOP green synthesis by use of organometallic selenium compound. *Current Applied Physics*, **10**, 553–556 (2010).
8. Jiang, Y., Gao, S., Li, Z., Jia, X., and Chen, Y. Cauliflower-like CuI nanostructures: Green synthesis and applications as catalyst and adsorbent. *Materials Science and Engineering B*, **176**, 1021–1027 (2011).
9. Prathna, T. C., Chandrasekaran, N., Raichur, A. M., and Mukherjee, A. Kinetic evolution studies of silver nanoparticles in a bio-based green synthesis process. *Colloids and Surfaces A: Physicochem. Eng. Aspects*, **377**, 212–216 (2011).
10. Bensebaa, F., Durand, C., Aouadou, A., Scoles, L., Du, X., Wang, D., and Le Page, Y. A new green synthesis method of $CuInS_2$ and $CuInSe_2$ nanoparticles and their integration into thin films. *Journal of Nanoparticle Research*, **12**(5), 18971903 (2009).
11. Sukirtha, R., Priyanka, K. M., Antony, J. J., Kamalakkannan, S., Thangam, R., Gunasekaran, P., Krishnan, M., and Achiraman, S. Cytotoxic effect of Green synthesized silver nanoparticles using Meliaazedarach against in vitro HeLa cell lines and lymphoma mice model. *Process Biochemistry*, **47**, 273–279 (2012).
12. Darroudi, M., Ahmad, M. B., Abdullah, A. H., Ibrahim, N. A., and Shameli, K. Effect of Accelerator in Green Synthesis of Silver Nanoparticles. *Int. J. Mol. Sci.*, **11**, 38983905 (2010).
13. Bonde, S. A biogenic approach for green synthesis of silver nanoparticles using extract of Foeniculumvulgare and its activity against *Staphylococcus aureus* and *Escherichia coli*. *Bioscience*, **3**(2), 5963 (2011).
14. Narayanan, K. B. and Sakthivel, N. Facile green synthesis of gold nanostructures by NADPH-dependent enzyme from the extract of Sclerotiumrolfsii. *Colloids and Surfaces A: Physicochem. Eng. Aspects*, **380**, 156–161 (2011).
15. Bo, L., Yang, W., Chen, M., Gao, J., and Xue, Q. A Simple and Green Synthesis of Polymer Based Silver Colloids and Their Antibacterial Properties. *Chemistry and Biodiversity*, **6**, 111116 (2009).
16. Hebeish, A. A., El-Rafie, M. H., Abdel-Mohdy, F. A., Abdel-Halim, E. S., and Emam, H. E. Carboxymethyl cellulose for green synthesis and stabilization of silver nanoparticles. *Carbohydrate Polymers*, **82**, 933–941 (2010).
17. Venkatpurwar, V. and Pokharkar, V. Green synthesis of silver nanoparticles using marine polysaccharide: Study of *in vitro* antibacterial activity. *Materials Letters*, **65**, 999–1002 (2011).
18. Zhang, X., An, C., Wang, S., Wang, Z., and Xia, D. Green synthesis of metal sulfide nanocrystals through a general composite surfactants aided solvothermal process. *Journal of Crystal Growth*, **311**, 3775–3780 (2009).

19. Leonard, K., Ahmmad, B., Okamura, H., and Kurawaki, J. *In situ* green synthesis of biocompatible ginseng capped gold nanoparticles with remarkable stability. *Colloids and Surfaces B: Biointerfaces*, **82**, 391–396 (2011).
20. Chen, J., Wang, J., Zhang, X., and Jin, Y. Microwave-assisted green synthesis of silver nanoparticles by carboxymethyl cellulose sodium and silver nitrate. *Materials Chemistry and Physics*, **108**, 421–424 (2008).
21. Ahmad, M. B., Tay, M. Y., Shameli, K., Hussein, M. Z., and Lim, J. J. Green Synthesis and Characterization of Silver/Chitosan/Polyethylene Glycol Nanocomposites without any Reducing Agent. *Int. J. Mol. Sci.*, **12**, 48724884 (2011).
22. Yang, G., Xie, J., Deng, Y., Bian, Y., and Hong, F. Hydrothermal synthesis of bacterial cellulose/AgNPs composite: A "green" route for antibacterial application. *Carbohydrate Polymers*, **87**, 2482–2487 (2012).
23. Wang, J., Gao, Z., Li, Z., Wang, B., Yan, Y., Liu, Q., Mann, T., Zhang, M., and Jiang, Z. Green synthesis of graphene nanosheets/ZnO composites andelectrochemical properties. *Journal of Solid State Chemistry*, **184**, 1421–1427 (2011).
24. Gao, S., Jia, X., Yang, S., Li, Z., and Jiang, K. Hierarchical Ag/ZnO micro/nanostructure: Green synthesis and enhanced photocatalytic performance. *Journal of Solid State Chemistry*, **184**, 764–769 (2011).
25. Young, J. L. and DeSimone, J. M. A special topic issue on green chemistry.Frontiers in green chemistry utilizing carbondioxide for polymer synthesis and applications. *Pure Appl. Chem.*, **72**(7), 1357–1363 (2000).
26. Wood, C. D., Cooper, A. I., and De Simone, J. M. Green synthesis of polymers using supercritical carbon dioxide. *Current Opinion in Solid State and Materials Science*, **8**, 325–331 (2004).
27. Kamrupi, I. R., Pokhrel, B., Kalita, A., Boruah, M., DoluiS, K., and Boruah, R. Synthesis of MacroporousPolymer Particles by Suspension Polymerization Using Supercritical Carbon Dioxide as a Pressure-Adjustable Porogen. *Advances in Polymer Technology*, 1–9 (2011).
28. Liu, Z. and Erhan, S. Z. "Green" composites and nanocomposites from soybean oil. *Materials Science and Engineering A*, **483–484**, 708–711 (2008).
29. Puskas, J. E., Seo, K. S., and Sen, M. Y. Green polymer chemistry: Precision synthesis of novel multifunctionalpoly(ethylene glycol)s using enzymatic catalysis. *European Polymer Journal*, **47**, 524–534 (2011).
30. Goretzki, C., Krlej, A., Steffens, C., and Ritter, H. Green Polymer Chemistry: Microwave-AssistedSingle-Step Synthesis of Various (Meth)acrylamidesand Poly(meth)acrylamides Directly from(Meth)acrylic *Acid and Amines. Macromol. Rapid Commun.*, **25**, 513–516 (2004).
31. Buffet, J. C. and Okuda, J. Group 4 Metal Initiators for the Controlled Stereoselective Polymerization of Lactide Monomers. *Chem. Commun.*, 4796–4798 (2011).
32. Pham, L. Q., Nguyen, H. V., Haldorai, Y., Zong, T., Johari, S., and Shim, J. J. Green Synthesis of Poly(vinyl pivalate). In *Supercritical Carbon Dioxide for the Preparation of Syndiotactic-rich Poly(vinyl alcohol)*. The 13th Asia Pacific Confederation of Chemical Engineering Congress APCChE (2010).
33. Iliescu, S., Plesu, N., Popa, A., Macarie, L., and Ilia, G. Green synthesis of polymers containing phosphorus in the main chain. *C. R. Chimie*, **14**, 647–651 (2011).
34. Yadav, J. S., Reddy, B. V. S., Sridhar Reddy, M., and Niranjan, N. Eco-friendly heterogeneous solid acids as novel and recyclable catalysts in ionic medium for tetrahydropyranols. *Journal of Molecular Catalysis A: Chemical*, **210**, 99–103 (2004).
35. Ma, J. and Hong, X. Application of ionic liquids in organic pollutants control. *Journal of Environmental Management*, **99**, 104109 (2012).
36. Brogle, R., Rys, P., and Rochat, E. Environmentally Friendly Recycling of Hexal in Medium Caliber Ammunition in Industrial Scale. *Propellants, Explosives, Pyrotechnics*, **25**(3), 153–157 (2000).
37. Sasai, R., Kubo, H., Kamiya, M., and Itoh, A. Development of an Eco-FriendlyMaterial Recycling Process forSpent Lead Glass Using aMechanochemical Process andNa_2 EDTA Reagent. *Environ. Sci. Technol.*, **42**, 4159–4164 (2008).

38. Patel, M., Thienen, N., Jochem, E., and Worrel, E. Recycling of plastics in Germany. *Resources, Conservation and Recycling*, **29**, 65–90 (2000).
39. Xue, M., Li, J., and Xu, Z. Environmental Friendly Crush-Magnetic Separation Technology for-Recycling Metal-Plated Plastics from End-of-Life Vehicles. *Environmental Science & Technology*, dx.doi.org/10.1021/es202886a (2012).
40. Scott, Gerald. Green polymers. *Polymer Degradation and Stability*, **68**(1), 17 (2000).
41. Carné Sánchez, A. and Collinson, S. R. The selective recycling of mixed plastic waste of polylactic acid and polyethylene terephthalate by control of process conditions. *European Polymer Journal*, **47**, 1970–1976 (2011).
42. Sato, O., Osada, M., Shirai, M., and Arai, K. Depolymerization of polyesters in sub and supercritical water. *10th European Meeting on Supercritical Fluids, Reactions, Materials and Natural Products Processing Proceedings* (2005).
43. Morin, C., Loppinet-Serani, A., Cansell, F., and Aymonier, C. Near and supercritical solvolysis of carbon fiber reinforced polymers (CFRPs) for recycling carbon fibers as a valuable resource: State of the art. *J. of Supercritical Fluids (available online)* (2012).
44. Lateef, H., Grimes, S. M., Morton, R., and Mehta, L. Extraction of components of composite materials: Ionic liquids in the extraction of flame retardants from plastics. *Journal of Chemical Technology and Biotechnology*, **83**, 541545 (2008).
45. Liu, F., Li, L., Yu, S., Lv, Z., and Ge, X. Methanolysis of polycarbonate catalyzed by ionic liquid [Bmim][Ac]. *Journal of Hazardous Materials*, **189**, 249–254 (2011).

Index

B

D

E

H

I

S